Introductory Semiconductor Device Physics

Introductory Semiconductor Device Physics

GREG PARKER

BSc, PhD, CPhys, MInstP, MIEEE
Department of Electronics and Computer Science
Microelectronics Group
The University of Southampton

PRENTICE HALL

New York London Toronto Sydney Tokyo Singapore

First published 1994 by
Prentice Hall International (UK) Limited
Campus 400, Maylands Avenue
Hemel Hempstead
Hertfordshire, HP2 7EZ
A division of
Simon & Schuster International Group

Typeset in 10/12 pt Times by
Columns Design and Production Services Limited

Printed and bound in Great Britain by
Redwood Books, Trowbridge, Wiltshire

Library of Congress cataloging-in-publication data
Parker, G. J.
Introductory semiconductor device physics / G.J. Parker.
p.   cm.
Includes index.
ISBN 0–13–143777–1 (pbk.)
1. Semiconductors.   I. Title.
QC611.P37   1994   93–24965
537.6′22—dc20   CIP

British Library cataloguing in publication data
A catalogue record for this book is available from the British Library

ISBN 0–13–143777–1 (pbk)

1   2   3   4   5   98   97   96   95   94

Contents

Preface

This book is an *introductory text* explaining the underlying physics of semiconductor devices. A number of similar books already exist on the market. Clearly, this book must differ *significantly* from those, in several ways, otherwise publication would not have been possible in the present economic climate.

Without referring to any *specific* texts, this work differs from the 'standards' in the following ways. There is no discussion of vacuum devices (valves, klystrons, magnetrons, etc.), with the exception of a brief look at the new microelectronic version. No unnecessary physics is introduced; all physical concepts discussed will be used. There are no problem sets at the ends of chapters. I recall with distaste my student days when there were only answers to odd-numbered problems. Worse still was when your answer did not tally with the one in the back of the book! Were you wrong, or did the author make a mistake? My way of preventing this dilemma is to solve all problems that come up fully within the text. Nobody is infallible, so if I've made an arithmetic error, it should be obvious by following the argument with a pocket calculator at the ready.

I do believe that it is necessary to have seen just a little quantum mechanics. The application of just one result of this topic area, namely the particle-in-a-box, allows one to have an insight into the workings of the new low-dimensional structures (LDSs) and the quantum-well lasers.

The 'old' physics, and the 'old' experiments are not reproduced. There's quite enough to learn if you just deal with the up-to-date material, but this does not mean that the old work is redundant. For a full understanding you *must* go back and see why the train of thought in a particular topic area followed the route it did. You may be surprised at the results of such an investigation. Quite often, the routes taken were not necessarily logical, in other cases certain routes were *forced* due to the current technology. Present-day technology may require a reassessment of old ideas, a specific example being the microelectronic valve already mentioned. *Always* question and *do not* take for granted! Pick up any three books on this subject and you will not only get three different views, you may also get three conflicting stories. If the stories are all the same, beware! The authors may have each read the same books and put the argument in the same way. They may all be right, or, conversely, . . . !!

A very obvious omission appears when reading the 'standard' texts on the p–n junction. It's this. How come the band diagrams *look inconsistent* (beware, looks can be deceptive) depending on whether you consider a p–n junction or an n–p junction (integrate through the p-region first or the n-region first)?? The chapter in question will make it clear what I'm talking about, but the answer is *not* obvious and it needs to be spelt out. This *sort* of problem can be a real stumbling block to a student just starting to study the subject. I hope I've answered that particular question and covered other similarly disturbing omissions from the alternative texts.

This is a *basic-level* introductory text, and as such it does not pretend to compete with *higher-level* texts such as Bar-Lev and Streetman (see Reading List). Bar-Lev and Streetman are excellent 'follow-on' texts to this one; they go much further into the subject, and do so with more mathematical rigour.

So, who is this book for? My main aim is to provide a *bridging text* between final year high-school level ('A'-level in the UK) and the first year of a university electronics or physics course. The baseline has to be drawn somewhere in order that every physical point is not taken back to first principles. That baseline is final year high-school level or its equivalent. I hope the writing style is readable, user-friendly and not too dry. I also hope it's not flippant or completely unintelligible. Finding a balance between all these and keeping your attention is not at all easy. The style *is* narrative and should read as if someone is actually talking to you. This means that the grammar is questionable in places. Apologies to the purists among you.

The construction of this book is minimalist. By that I mean that I have tried to introduce the minimal amount of physics for an understanding of the topic areas to be covered. I hope that I have not introduced any physical principles or concepts that are not actually used in the following chapters. This minimalist approach also applies to the order of the chapters. If you have the staying power to read the whole book you should find that one chapter leads logically to the following chapter. The idea is that just enough material is covered to get started on the following chapter. This being the case I will briefly run through the book layout.

Chapter 1 discusses the atoms that go into making semiconductor materials and how they bond in the solid. I believe it is necessary to see the alternative bonding mechanisms in order to appreciate the uniqueness of the covalent bond. The understanding of how atoms bond together leads directly to the Chapter 2 material on energy bands and effective mass. The energy-band approach is used throughout the book as it gives a good physical *picture* of how devices work, and I much prefer physical pictures to mathematical equations. For this reason energy bands are introduced early on, and they are discussed at some length. The effective-mass concept leads directly to a deeper understanding of holes in semiconductors and introduces energy–momentum diagrams. Energy–momentum diagrams give a good pictorial understanding of why some materials are efficient photon producers, and others are not (except for the case of porous silicon!).

With the differences between metals, insulators and semiconductors having been covered, Chapter 3 then concentrates on just the semiconductor class of solids, and how the number density of charge carriers in them can be calculated. With a knowledge of the number densities of free carriers in semiconductors, all that is required is an introduction to the mechanisms that move these carriers, in order to calculate the currents flowing in real semiconductor devices. Chapter 4 introduces these mechanisms, specifically the drift and diffusion of carriers. In many ways Chapter 4 is the 'backbone' of the book, not only because it is so fundamental to semiconductor device physics, but also because a good understanding of this work allows you to make a start on designing your own devices (via the energy-band diagrams).

Chapter 5 introduces our first real semiconductor device, the Gunn diode. Why this device and not say the p–n junction? The Gunn diode structure is much simpler, and only carrier drift is considered. In addition, there are no junctions with dissimilar materials on either side of the junction, and bipolar conduction does not have to be considered. For these reasons the Gunn diode is discussed before the most common two-terminal device, the p–n junction diode of Chapter 6.

I believe that the p–n junction is a pretty difficult device to come to grips with. Many people picking up this book may think that they already know quite a bit about p–n junctions, but it is in fact quite a complicated device! Since the light emitting diode (LED) is itself a p–n junction, Chapter 7 naturally follows on from the p–n junction study. The p–n junction can also detect photons as well as emit them so photodiodes are also included in Chapter 7.

Chapter 8 discusses the fundamental active bipolar device, the bipolar transistor. Looking back on the previous chapters it should be clear how they have been hung together in order to get to grips with the bipolar transistor quickly. In the flow of this book, with one chapter leading logically to the next, Chapter 9 which discusses field-effect transistors, is perhaps the odd one out. In studying the metal–oxide semiconductor field-effect transistor (MOSFET), for example, it is difficult to see how one can lead progressively from simpler devices to this one, since the simpler device (the MOS capacitor) is already pretty difficult to understand. For this reason the field-effect transistor chapter is self-contained, and is presented in a less rigorous form than in the 'standard' texts.

Chapter 10, the semiconductor laser, follows logically from Chapter 8 in the same way that Chapter 7 (the LED) followed logically from Chapter 6 (the p–n junction). Why? Because the double-heterojunction laser can also be a double-heterojunction bipolar transistor (and has been used as such in optoelectronic integrated circuits!).

This leaves the heaviest (and I believe the most interesting) physics to Chapter 11, the subject being quantum mechanics. You may get a little annoyed that throughout the book you find that in order to *fully* understand a device mechanism quantum mechanics is necessary, and I've gone and stuck this chapter at the end. Why? Well it's true, a lot of quantum mechanics is needed for a full understanding of devices, especially the newer ones such as quantum-well lasers, resonant tunnelling devices and the low-dimensional structures. It's also needed for the simpler devices we come up against at the beginning of the book. The reason that this chapter is at the end of the book is that I have tried, in lectures, the alternative approach of introducing quantum mechanics right at the beginning (where I believe it belongs). The exercise was a dismal failure. The shock of the new was too great and made the group believe that the *whole* subject was incredibly difficult, which of course it isn't. (Quantum mechanics isn't too difficult either if you don't think too much about what you're doing!). So, in this book, just as in later lecture courses, I've chickened out and the quantum mechanics all comes at the end. However, once you've gone through this last chapter, look at the earlier ones again, and see how this fascinating subject applies to all that we've covered before.

I am just left with the obligatory, but nevertheless pleasant, task of mentioning those who have helped me along the way.

First of all I must thank Dr Peter Ashburn, who not only read a lot of the text, but also helped in points of detail, and acted as my inspiration in view of his own recently published book on bipolar transistors. Many thanks also go to Dr Ian Post and Dr Alan Shafi for numerous rescues from my entanglements with the word processor. I would also like to thank Mr Jim Smith whose extensive physics knowledge-base was plundered on many occasions. Much warm appreciation goes to Professor Henri Kemhadjian, Head of the Microelectronics Group at the University of Southampton for providing the stimulating environment which made the writing of this book a possibility.

Finally, thank you Dave Squibb of Tavistock, Devon, for starting it all off!

GREG PARKER
Brockenhurst, Hampshire, England

Symbols

A	area (of device)		transistor)
C_{DEP}	depletion capacitance	I_C	collector current (bipolar
C_{DIFF}	diffusion capacitance		transistor)
C_j	junction capacitance	I_d	drift current
C_{OX}	oxide capacitance	I_D	drain to source current in
C'_{OX}	oxide capacitance per area		FET
c	velocity of light *in-vacuo*	$I_{D(SAT)}$	drain saturation current
D_n, D_p	electron, hole diffusion	I_F	forward current
	coefficient	$I_{n,p}$	electron, hole current
d	atomic spacing	I_0	reverse saturation current
dE	incremental energy	I_P	photocurrent
d_{OX}	oxide thickness	I_{PRIM}	primary photocurrent
E	energy	I_{SC}	short-circuit current
E_C	conduction band-edge energy	J	current density
E_F	Fermi-level energy	$J_{D(SAT)}$	drain saturation current
E_g	band-gap energy		density
E_i	intrinsic Fermi-level energy	J_n, J_p	electron, hole current density
E_V	valence band-edge energy	J_{TOTAL}	total current density
$\mathbf{E}$	electric field	K	Kelvin
E_m	maximum electric field	k	Boltzmann's constant
$\mathbf{F}$	force	k	propagation constant
F	flux	L	gate length
f_T	maximum frequency of	L	Quantum-well width
	operation	L_n, L_p	electron, hole diffusion length
G	generation rate	m	mass
g_D	channel conductance	m_n^*, m_p^*	effective mass of electron,
$g_{m(SAT)}$	saturated transconductance		hole
h	Planck's constant	m_0	rest mass of electron
$\hbar$	reduced Planck constant $h/2\pi$	N	Avogadro's number
I	current	N_A	concentration of acceptors
I_B	base current (bipolar	N_C	effective conduction band

	density of states	r_e	emitter resistance
N_D	concentration of donors	T	temperature
N_V	effective valence band density of states	T	time period
		t	time
n	electron concentration	t_{tr}	transit time
n_i	intrinsic carrier concentration	V	applied bias, volts
n_n, n_p	electron concentration in n-type, p-type material	V_{BE}	base–emitter voltage (bipolar transistor)
n_{n0}	equilibrium concentration of electrons in n-type material	V_{bi}	built-in voltage, junction potential
n_{p0}	equilibrium concentration of electrons in p-type material	V_{CE}	collector–emitter voltage (bipolar transistor)
n	Principal quantum number, integer	V_D	drain voltage (FET)
		V_{DEP}	depletion potential
P	pressure	V_F	forward bias voltage
P_{fd}	Fermi–Dirac function	V_G	gate voltage (FET)
P_{OPT}	optical power	V_{INV}	strong inversion potential
p	momentum	V_{OC}	open-circuit voltage
p	hole concentration	V_n	electron electrostatic potential
p_n, p_p	hole concentration in n- and p-type material	V_p	hole electrostatic potential
p_{n0}	equilibrium hole concentration in n-type material	V_{po}	pinch-off voltage
		V_R	reverse bias voltage
		V_T	threshold voltage
p_{p0}	equilibrium hole concentration in p-type material	V	potential energy
		v_d	domain velocity
Q	space-charge	v_d	drift velocity
Q_{DEP}	depletion-layer charge	$v_{d(sat)}$	saturation drift velocity
Q_M	charge on gate electrode (FET)	v_{th}	thermal velocity
Q_n	inversion-layer charge (MOSFET)	W	total depletion-layer width
		W_B	effective base-width
		W_M	maximum depletion-layer width
q	elementary charge	x, y, z	space coordinates
R	resistance	x_p, x_n	depletion width p-side, n-side
R_L	load resistance		

Greek and other symbols

α	absorption coefficient	σ	conductivity
α_0	small signal common-base-current gain	τ	lifetime
		τ_B	base minority carrier lifetime
α_T	base transport factor	τ_n, τ_p	electron, hole mean free time
β	common-emitter current gain	τ_{TR}	transit time
γ	emitter efficiency	τ_d	dielectric relaxation time
$\delta n, \delta p$	excess electron, hole concentration	ϕ	wave amplitude
		ψ	position-dependent wavefunction
ε_0	permittivity of free space		
ε_r	relative permittivity	ψ_p	electrostatic potential
ε_{Si}	permittivity of silicon	Ψ	time- and position-dependent wavefunction
ε_{OX}	oxide permittivity		
η	quantum efficiency	ω	angular frequency
θ	angle	Ω	ohm
$\varkappa$	scaling factor	$\Im$	imaginary part of complex number
λ	wavelength		
ν	frequency	$\Re$	real part of complex number
μ_0	permeability of free space	∇	vector differential operator
μ_n	electron mobility	∇^2	Laplacian operator
μ_p	hole mobility	$|x|$	absolute magnitude of x
ρ	resistivity	$\sqrt{x}$	square root of x
ρ_s	charge density		

CHAPTER ONE

Atoms and bonding

In order to understand the physics of semiconductor devices, we start by looking at how atoms bond together to form the solids that go into making the devices.

Recall that an atom is the smallest unit that has the properties of the element concerned. The atom is composed of a nucleus containing the protons and neutrons; surrounding the nucleus are the electrons. I use the word 'surrounding' rather than 'orbiting' as orbits imply nicely spherical or elliptical trajectories. Quantum mechanics (Chapter 11), which is our best description of matter on the atomic scale, tells us the *probability* of finding an electron about the nucleus but it does not describe well-defined electron trajectories. The probability mapping of an electron about a nucleus may in some instances have a spherical symmetry, but this does not mean that the actual electron trajectory is spherical. It means that we have a higher probability of finding the electron *on* the spherical shell than some distance away from it. Quantum mechanics doesn't say where *specifically* on the shell the electron is, so we can't follow it's motion, and we cannot draw trajectories.

What we do know is that atoms are sometimes able to combine with themselves or other atoms, and that it is the outermost or *valence* electrons that are the most important to consider in these circumstances. These *valence* electrons govern the *chemistry* of the atoms.

We also know that atoms may come together to form gases, liquids or solids depending on the strength of the attractive forces between them. Several types of atomic bonding have been identified, including *ionic, covalent, van der Waals, hydrogen*, and *metallic*. Whatever the name given to the type of bonding, in all cases it is the *electrostatic force* acting between charged particles that is responsible for all the forms of bonding.

1.1 The periodic table

Figure 1.1 reproduces part of the periodic table. This is not a chemistry book, which in some ways is a little unfortunate since the periodic table makes a fascinating study in itself. Dmitri Mendeleev first attempted to arrange the elements according to periodic similarities. He put similar elements in the same *vertical* column. This was a brave move. Note that the periodic table *is not* simply a list of the elements in increasing atomic mass. Tellurium has a higher atomic mass than iodine, yet is listed *before* iodine in the table since elements in the same vertical *group* must be similar. The monotonic atomic mass increase also breaks down elsewhere in the table (refer to a complete table and see if you can find where).

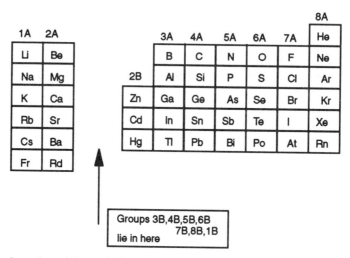

Figure 1.1   A section of the periodic table.

However, the main reason for introducing the periodic table here is that the next section on bonding in solids assumes some knowledge of the elemental groups. Referring to Figure 1.1 again, group 1A contains the *alkali metals* lithium (Li), sodium (Na), potassium (K), . . ., and these combine readily with the *halogens* of group 7A fluorine (F), chlorine (Cl), bromine (Br), . . ., producing *ionic solids* such as sodium chloride (NaCl), potassium chloride (KCl), etc.

Group 8A contains the noble gases helium (He), neon (Ne), argon (Ar), ... These gases have a full complement of valence (outer lying) electrons and do not readily combine with other elements. However, the rare (or noble) gases are not called the inert gases anymore because compounds containing xenon (Xe), krypton (Kr) and radon (Ra) *can* be formed.

The *elemental semiconductors* silicon (Si) and germanium (Ge) are to be found in group 4A (certain forms of tin (Sn) and carbon (C) are also semiconducting).

The *compound semiconductors* can be formed from groups 3A and 5A, giving the well-known 3–5 (III–V) compounds gallium phosphide (GaP), gallium arsenide (GaAs), AlGaAs, InGaAsP, . . ., etc. They can also be formed from groups 2B and 6A giving the 2–6 (II–VI) compounds zinc sulphide (ZnS), zinc telluride (ZnTe), cadmium sulphide (CdS), mercury cadmium telluride (HgCdTe), ... Since this is a book discussing *electronic devices* the groups 2B–6A are the ones we'll be most concerned with.

In Section 3.1 we shall see that elements with one valence electron *more* than elemental silicon can be incorporated into silicon, making it more conducting. This being so we will be interested in the group 5A elements phosphorus (P), and arsenic (As), for modifying the conductivity of silicon. We can also modify the conductivity of silicon by incorporating an element with one valence electron *less* than that of silicon, i.e. group 3A elements. We shall see that boron (B) is the element usually used in this case.

1.2 Ionic bonding

Atoms with low ionization energies (e.g. group 1 elements) lose electrons easily to atoms with high electron affinities (e.g. group 7 elements). This process leads to electron transfer and the formation of charged *ions*, a positively charged ion for the atom that has lost the electron and a negatively charged ion for the atom that has gained an electron. The ions formed have a noble-gas configuration of valence electrons, i.e. complete shells. The *ionic crystal* itself is formed by the ions coming together in an equilibrium configuration. Figure 1.2 shows how such an equilibrium configuration can come about. There is an atomic separation, r_0, at which there is *balance* between the repulsive and attractive electrostatic forces. The repulsive forces are between the like-charged electrons surrounding the atoms, and, of course, the like-charged atomic nuclei. The attractive forces are between the *ions* that have been formed by the electron transfer. Typical ionic

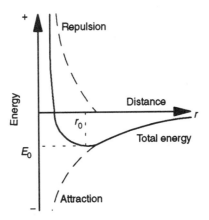

Figure 1.2 Equilibrium in a crystal lattice.

crystals can be formed by elements combining from groups 1 and 7 of the periodic table, e.g. sodium chloride (NaCl) and caesium chloride (CsCl). The strength of the ionic bond is reflected in the crystal typically having a high melting point and being hard. Although hard, ionic crystals tend to be brittle as their structure does not allow for much movement between atoms since they are well-ordered. Because ionic crystals are formed by charge-separated atoms they can often be dissolved by charge-separated (polar) liquids such as water.

1.3 Covalent bonding

Covalent bonding is of the most interest to us since the elemental semiconductors are bonded by this mechanism. The strongly directional covalent bond is favoured by elements in the central part (group 4) of the periodic table. These atoms prefer *electron sharing* in their bonding arrangement as this is more energetically favourable than the *electron transfer* of the ionic solids. Silicon (Si), germanium (Ge) and diamond (C) are examples of solids that are purely covalent in character. They are also capable of being made into semiconductor devices (including diamond). The strength of the covalent bond is shown by the hardness and high melting point of the crystals; this is clearly exhibited in diamond and silicon carbide, with diamond having the greatest hardness of any material. Unlike the ionic solids the covalent solids are insoluble in all ordinary liquids, which accounts for some of the very 'nasty' chemicals encountered in the semiconductor processing industry.

As we move away from the purely covalent and purely ionic solids it should come as no surprise that the compounds we encounter exhibit a mixture of both ionic and covalent bonding. The important 3–5 *compound semiconductors* (named because they are mixtures of group 3 and group 5 elements) such as gallium arsenide (GaAs), indium phosphide (InP), aluminium gallium arsenide (AlGaAs) and gallium indium arsenide phosphide (GaInAsP) all show mixed covalent/ionic bonding. As we move away from the central part of the periodic table the compounds become slightly more ionic in character. For example, the 2–6 compound semiconductors such as zinc selenide (ZnSe) and cadmium telluride (CdTe) are more ionic in nature than the 3–5 compound semiconductors just mentioned. These compound semiconductors are vitally important in manufacturing optoelectronic devices such as LASERs and LEDs because (unless we resort to quantum wire structures) the elemental semiconductors (Si, Ge) are extremely inefficient at producing photons.

1.4 Metallic bonding

Metals, such as copper, have one valence electron. The valence electrons of nearby copper atoms are shared in a similar way to covalently bonded solids but it is not possible for seven close copper neighbours to come together to gain a noble-gas configuration of valence electrons as occurs in the covalent bond. Instead, the electron sharing that occurs in metals is not as restricted as in the covalent solid. In a metal, such as copper, the atoms readily lose the single electron (become ionized) and the positive copper ions reside in a sea of free valency electrons called a *free electron gas*. The overall bonding in a metallic crystal is due to the strong electrostatic attraction between the ions and the free electrons. As the free electrons are very mobile, metals have a high electrical conductivity (we shall see how band-theory accounts for high electrical conductivity in the next chapter). The metallic bond is weaker than either the ionic or covalent bond, which is reflected in the fact that metals are easily deformed (malleable and ductile).

1.5   van der Waals bonding

We saw that the stability of ionic and covalent solids came about by their acquiring a noble-gas configuration of valence electrons, either by transferring electrons, or by sharing electrons. This being the case, what do the noble gases themselves do? The noble gases such as helium (He), argon (Ar), and neon (Ne) do show weak, short-range attractive electrostatic forces. They must show some sort of attractive force to one another otherwise they could not be liquified or solidified, which in fact they can, despite the fact that no ionic, covalent or metallic bonding is possible. In this instance it is the van der Waals forces that are responsible. Although the valence electron distribution around the noble-gas atoms is symmetrical on the average (i.e. they are non-polar showing no charge separation), at any instant the distribution is *not* symmetrical and the atom *does* show a charge separation (it possesses an electric dipole moment). This charge separation is the origin of the van der Waals force. As such it is a very weak, short-range force, dropping off as r^{-7} where r is the atomic separation. Such a weak bonding force naturally results in low melting and boiling point solids with little mechanical strength.

Although the strengths of the bonding forces give rise to the possibility of the formation of gases, liquids or solids, for semiconductor physics studies we are most interested in solids. The solids themselves may be crystalline, with atoms in well-defined positions on a crystal *lattice*, or they may be polycrystalline or even amorphous. Polycrystalline solids are, as the name suggests, made up of many smaller crystalline regions, and amorphous solids have no long-range crystal structure at all, hence the word 'amorphous' (without form). For a full understanding of today's semiconductor devices we need to know the properties of all these forms of solid. Silicon is used in crystalline and polycrystalline forms in the fabrication of integrated circuits. Polysilicon (polycrystalline silicon) is used extensively to interconnect different device areas into one large circuit; it is thus used as an ultrafine wire to electrically connect all the devices on a chip.

Amorphous silicon can be used as the starting point for producing either polysilicon or crystalline silicon, or it can be used as it is in solar cells. Amorphous solids include such materials as glass, pitch and many plastics. Because the bond strengths between molecules vary in amorphous solids there is no sharply defined melting point.

This brief resume of bonding in solids is all I want to cover on that particular subject, but we do need a result from quantum mechanics in order to see how solids can form into semiconductors, insulators or metals. Quantum mechanics is introduced more fully in Chapter 11 but the result we want to use now is called the Pauli Exclusion Principle. Electrons are *fermions* (spin-half particles) and as such they have to follow certain rules (*statistics*) that govern the behaviour of spin-half particles. These are known as Fermi–Dirac statistics after the people who developed them. The rule we wish to use is the fact that no two electrons in the same system (including systems where atoms come together to form solids) can be described by the same set of parameters. Put another way, these parameters are called quantum numbers; but all that concerns us for the moment is that we can only have a maximum of two electrons in a given energy-level, and then we can only have two if they have different spins (spin-up and spin-down). You can see trouble brewing once we start to bring together large numbers of atoms. Each individual atom has electrons in similar energy-levels but once we start bringing the atoms close together the Pauli exclusion principle does not allow all these electrons, with similar quantum numbers, to come together. Something must 'give' and this is discussed in the next chapter.

Before leaving this introductory section it is worth mentioning that there are statistics other than Fermi–Dirac, although since we are going to be dealing mainly with electrons (and holes) these are the statistics of importance to us. The other statistics are called Maxwell–Boltzmann, and Bose–Einstein, again after the men who researched them. Bose–Einstein statistics apply to particles with zero or integer spin called *bosons*. Photons (light particles) are examples of spin-zero particles. Bose–Einstein particles differ significantly from spin-half particles in that they actually *like* to all be in the same (lowest) energy-level. It is this behaviour that is, in part, responsible for the phenomenon of *superconductivity* that occurs in certain materials at low temperatures. Remember, only *two* fermions (spin-half particles) like to be in the same energy-level.

Maxwell–Boltzmann statistics apply to widely separated particles of any spin and as such are typically applied to *ideal gas* systems. However, an important result that we shall be using is that the more complicated Fermi–Dirac equations simplify to the Maxwell–Boltzmann equations under certain conditions.

1.6 Start a database

The elemental semiconductor around which most of the electronics industry is based is of course silicon. Create your own database by writing down as much information as you can find on silicon *and* its compounds. To start, it has fourteen electrons and an atomic mass of 28.086. It therefore has fourteen protons and fourteen neutrons in the nucleus. The outermost *valence* shell has four electrons in it, therefore silicon will want to pair up with four nearest neighbours (to attain a noble-gas configuration) and this it does, crystallizing in the diamond lattice structure. The lattice constant (equilibrium distance between atoms) is 54.3 nm. The density of silicon is 2.33 g/cm^3 and its melting point is 1415 °C. Silicon readily forms a tough, chemically resistant oxide SiO_2, which is used extensively in the fabrication of devices and is the major property that still keeps silicon integrated circuit technology viable against other semiconductor material systems that seem to offer much better performance. These other material systems, which include gallium arsenide (GaAs) and indium phosphide (InP), would be far more suitable for certain device applications, but the lack of a tough, resistant naturally occurring, or native oxide hinders their widespread use considerably. There are, of course, device types that can *only* be made with these compound semiconductors, for instance certain microwave and optoelectronic devices. In these cases the device engineers create ingenious processing schedules in order to overcome the problems caused by the lack of a native oxide.

We shall learn a lot more about the properties of silicon, and how it is used in semiconductor devices, in the forthcoming chapters.

CHAPTER TWO

Energy bands and effective mass

2.1 Semiconductors, insulators and metals

The electrical properties of metals and insulators are well known to all of us. Metals tend to conduct electric current very well. This is why ordinary connector wire typically consists of a bundle of copper wires. Insulators, on the other hand, do not seem to conduct electric current at all, and they are therefore used in situations that take advantage of this property. We see the use of insulators in car sparking plugs where we want to keep the very high spark voltages away from the rest of the engine block, and we see them supporting high-tension electricity cables. Insulators tend to be very good at blocking electric current.

Although everyday experience has already taught us a lot about the electrical properties of metals and insulators, the same cannot be said about semiconductors. We can readily obtain samples of metal or insulator to experiment with in order to discover their electrical properties, but semiconductor material, such as germanium or silicon, is not quite so available to us. For this reason, most people would not know what would happen if you were to connect a battery across a piece of silicon. Would it conduct well? Or would it act like an insulator? Clearly, the name 'semiconductor' implies that it conducts somewhere between the two. This is not a great help to us when the difference in electrical conductivity between metals and insulators, that is, how well they conduct electricity, can be a factor of 10^{28} or more, that is, one followed by twenty eight zeros! The aim of this chapter is to try and find out why metals are good conductors of electric current, why insulators do not seem to conduct at all and how well (or badly) semiconductors conduct.

The approach we shall use in order to understand the very different behaviour of semiconductors, insulators, and metals is called the Band Theory of Solids. Don't be put off by the name, I've only included it so that if by the end of the chapter you wish to know more, you will know the subject heading we're talking about. Any good general physics text should discuss simple band theory.

The electrons surrounding a nucleus have certain well-defined energy-levels, and a theory called quantum mechanics allows us to actually calculate these energies, certainly for free atoms. We saw from Chapter 1 that electrons don't like to have the same energy in the same potential system, and that the most we could get together in the same energy-level was two, provided that they had opposite spins. This is called the Pauli exclusion principle and turns up again in quantum mechanics. So what happens when we start to bring lots of individual atoms together to form a solid? It looks like there's going to be a serious problem here! In an idealized thought experiment we can imagine bringing together two separate silicon atoms as the starting point in making a small piece of solid silicon. Silicon has fourteen orbiting electrons, each in its own specific energy-level. The two atoms we're trying to bring together with their fourteen electrons are fine while they're some distance apart, but what happens as we try to push them together as they would be in the solid? The atoms in solids are typically Angstroms apart where an Angstrom is 10^{-10} metre. In SI units we would say that atoms in a solid are typically a few tens of nanometres apart. At these very close distances we are going to run into trouble with the Pauli exclusion principle. The electrons of one atom are close enough to the electrons of the neighbouring atom to appear to be part of the same system. But the electrons of one atom are also sitting in the *same* energy-levels as the corresponding electrons in the neighbouring atom, so something has to give as we bring the atoms together. What happens is the energy-levels subdivide so that no more than two electrons occupy the same level as required by the Pauli principle. So for two atoms coming together, each distinct energy-level splits into two slightly different energy-levels. For four atoms coming together each energy-level splits into four energy-levels of slightly different value. The number of levels is clearly going to be the same as the number of atoms coming together to make the solid. This is a pretty big number! Let's calculate this number for silicon.

Calculation

Consider 1 cm^3 of silicon. How many atoms does this contain?

The atomic mass of silicon is 28.1 so 28.1 kg of silicon contains Avogadro's number of atoms, where Avogadro's number N is 6.023×10^{26} atoms/kmol.

The density of silicon is $2.3 \times 10^3 \text{ kg/m}^3$, so 1 cm^3 of silicon weighs 2.3 g and therefore contains $(6.023 \times 10^{26}/28.1 \times 10^3) \times 2.3 = 4.93 \times 10^{22}$ atoms.

This means that in a piece of silicon just one cubic centimetre in volume, each electron energy-level has split up into 4.93×10^{22} smaller levels! The difference in energy between each of these smaller levels is so tiny that it is more reasonable to consider each of these sets of smaller energy-levels as being continuous *bands* of energy, rather than considering the enormous number of discrete individual levels. This physical situation is shown in Figure 2.1. Notice that something interesting has happened in forming the solid. The discrete energy-levels have come together to form allowed energy bands where electrons may be found, and

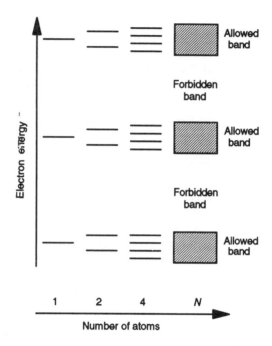

Figure 2.1   Energy-band formation in a solid.

between these allowed bands there are forbidden regions where no electrons are to be found. These forbidden regions are seen to form quite naturally from our simple picture of bringing together discrete atoms. What is a little difficult to understand is that electrons with energies *in* these regions are simply *not allowed*. A more detailed understanding of what is physically occurring in these forbidden regions requires some quantum mechanics, which we will study later.

If you feel unhappy about calling all these energy-levels a continuous energy band, then think of an experiment that will show the discrete nature of the energy-levels in a band. If there is no physical experiment that will demonstrate the discrete nature of a band, then it must be considered continuous.

Calculation

Calculate the energy-level separation between discrete levels in an energy band in a solid. Having calculated this energy, what are the consequences, regarding the temperature at which an experiment to determine the energy-level separation, must be carried out?

The energy spread of a band is of the order $1\,\text{eV}$ or less. The energy-level separation between levels is therefore approximately $1\,\text{eV}$/number of atoms in the solid $= 10^{-20}$ eV or less. This energy-level separation within the band is so small that the band appears to be continuous.

Boltzmann's constant in eV units is 8.625×10^{-5} eV/K. Simply taking $E=kT$ we solve to find $T = 1.16 \times 10^{-16}$ K. Any system we are looking at has to be at a temperature lower than this if we are going to see such tiny energy-level differences. Such a low temperature is indiscernible from absolute zero and the Laws of Thermodynamics don't allow us to reach absolute zero.

This understanding of the formation of allowed bands (where electrons are to be found) and forbidden bands (where electrons do not exist) when atoms come together as in a solid is all there is to actually understanding the basics of band theory. Also, the above description, with one additional piece of information, is all we need in order to understand the fundamental electrical differences between semiconductors, insulators and metals.

The additional piece of information is this. *A band that is completely full of electrons cannot partake in electrical conduction.* It is not surprising that a band that is completely *empty* of electrons cannot partake in electrical conduction since there aren't any electrons there to conduct! But what about the full band case? This is a little more complicated. If an electron is going to become a conduction electron, it will move when an electric field is applied. If the electron is moving due to the application of an electric field, then it has gained some energy from the field. If the electron has gained some energy from the field, then it has moved into a higher energy state (than it was in originally), so it must move higher into the energy band. In a completely full band this cannot occur because if an electron is at the top of the band there are no allowed states immediately above it. A forbidden band lies immediately above the top of an allowed band. If an electron is some way down in a full band, then it still can't gain energy from the electric field and move up within the band because all the allowed states above it are already filled, by definition of a full band. This is why we state that a full band cannot conduct.

An electron that *is* partaking in conduction will be in the *conduction* band. The conduction band is the first *allowed* band lying above the *valence* band. At zero degrees Kelvin all the electrons sit in the lowest allowed energy-levels. The *valence* band is the highest filled energy band at zero Kelvin. Figure 2.2 shows a possible physical situation.

We are now ready to understand the difference between semiconductors, insulators and metals.

2.2 Semiconductors

The simple energy diagram for a semiconductor at low temperature is shown in Figure 2.2. At very low temperatures the *valence* band is *full*, and the *conduction* band is *empty*. Recall that a full band cannot conduct, and neither can an empty

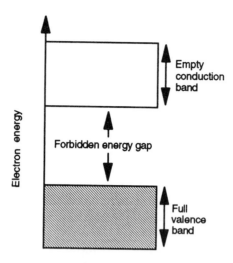

Figure 2.2 Semiconductor energy bands at low temperature.

band. At low temperatures, therefore, semiconductors do not conduct, they behave like insulators!

If energy is supplied to take an electron from the valence band, across the forbidden energy gap, up into the conduction band, then the electron that has made it to the conduction band is now available for conduction. In addition there is now a vacant electron energy state left in the valence band. This vacant state is called a *hole* and it behaves like a positive charge carrier with the same magnitude of charge as the electron, but of opposite sign. The band diagram for this situation is shown in Figure 2.3, where the conduction band is populated by electrons leaving a similar number of vacant states (*holes*) in the valence band. This is an important property of *intrinsic*, or undoped, semiconductors. The number of electrons in the conduction band must equal the number of holes in the valence band because each electron has given rise to one hole! For *extrinsic*, or doped, semiconductors where we can put an excess of either electrons or holes into the material, this is no longer true. The origin of holes will be made much clearer in the next chapter when we consider doping semiconductors, that is, putting other materials into semiconductors that drastically change the semiconductors' electrical properties. But for now bear in mind just two things. The hole in the valence band is able to carry current in just the same way as an electron in the conduction band, except that the hole has positive charge. The hole *is not* a free particle, it can only exist within a solid where there is an electron missing from an otherwise full band. A hole is simply a vacant electron state, some say a missing electron. You cannot make a hole gun in the same way as you can make an electron gun because a hole is not a free particle!

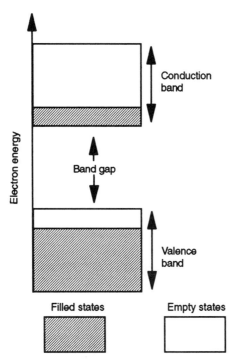

Figure 2.3   Semiconductor energy bands at room-temperature.

Consider Figure 2.4, which shows just the valence band of a semiconductor in more detail. The semiconductor is at a finite temperature so that some electrons have been thermally excited to the conduction band leaving behind empty states, *holes*, at the top of the valence band. Since the valence band is now no longer full, it may conduct electric current, and this is just what the holes do. When a material can conduct by electrons in the conduction band and by holes in the valence band we say the material is a *bipolar* (two-carrier) conductor.

Note that increasing electron energy is in the positive y-direction. We must supply energy to electrons in order for them to cross the forbidden energy gap. By contrast, hole energy increases downwards (a little thought will show that this is reasonable) and the more energetic holes are to be found lower down in the valence band.

The difference in energy between the top of the valence band and the bottom of the conduction band is called the energy gap or the *band gap* of the semiconductor. This is perhaps the most important single parameter describing semiconductor behaviour and it tells us a great deal about the semiconductor. Silicon has a band gap of about 1.1 eV at room-temperature (semiconductor band gaps usually vary just a little with temperature) so an electron must gain 1.1 eV of energy to cross the band gap. What excitation mechanisms can do this? Thermal

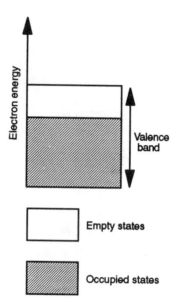

Figure 2.4   Valence band of a semiconductor.

energy can promote the electrons but as the thermal energy available at room-temperature is only 25 meV (from $E = kT$), only very few electrons will be thermally excited to the conduction band at room-temperature. In case you think that *no* electrons should reach the conduction band if the gap is 1.1 eV and only 25 meV is available, then we will have to go a little deeper into the theory and say that the excitation rate is proportional to an exponential function:

$$EXCITATION\ RATE = const. \times \exp\left(-\frac{GAP\ ENERGY}{kT}\right) \quad (2.1)$$

so, although the rate is very low at room-temperature, it is a strong function of temperature and increases exponentially with increasing temperature.

Can an electric field promote an electron from the valence band to the conduction band? Let me give the answer first before we calculate it. For the common semiconductor materials (Si, GaAs) and for low electric fields, electrons *cannot* be promoted across the band gap.

Calculation

A typical mean free path of an electron in a solid is of order 10 nm. The mean free path is the distance the electron travels before it suffers a collision and comes to a stop before carrying on. So for an electric field of 10^8 V/m an electron would gain 1 eV from the electric field between collisions. This is an enormous electric field required for such a small gain in electron energy and for this reason excitation by this mechanism is not the normal way of promoting electrons.

Having said this, very high electric fields which can promote electrons *do* exist in some devices, and some devices make use of them (e.g. avalanche photodiodes). In most other circumstances such effects are usually detrimental and can destroy the device.

The final excitation mechanism that I wish to consider is electromagnetic radiation or *photons*. A simple equation relates the energy of a photon to its wavelength in microns (10^{-6} m or μm):

$$ENERGY \text{ (in eV)} = 1.24/WAVELENGTH \text{ (in microns)} \qquad (2.2)$$

Using equation (2.2) we see that photons of wavelength 1.1 μm (or less) will promote electrons across the silicon band gap. Photons in this wavelength regime are just in the infrared part of the spectrum (near infrared) so, energetically speaking, they are rather weak. We can see that visible wavelength photons, which are more energetic than infrared photons, are going to be important in our consideration of conduction in semiconductors such as silicon. Because visible photons are effective in enhancing conduction in semiconductors, semiconducting materials are often used as efficient *photodetectors*.

What about the converse situation? Can an electron from the conduction band recombine with a hole in the valence band and produce a photon of energy E_g? Under certain conditions the answer is yes! This is the basis of operation for LEDs and LASERs but it requires *direct band-gap* semiconductors such as gallium arsenide (GaAs) or indium phosphide (InP).

2.3 Insulators

Figure 2.5 shows an energy diagram for an insulator. The main characteristic of an insulator, as can be seen in the diagram, is that the full valence band lies well below an empty conduction band. Since the valence band is full the electrons in it will not conduct electric current, and the same applies, of course, to the empty conduction band. The energy gap is the difference in energy between the top of the valence band and the bottom of the conduction band. This is the energy that must be supplied to an electron so that it can cross the forbidden energy gap to become a conduction electron. It is the magnitude of this gap that differentiates between semiconductors, insulators and metals.

If an insulator is to remain insulating at room-temperature then the size of the forbidden energy gap, or simply the energy gap, must be much greater than the thermal energy that an electron can gain at room-temperature. Since an insulator band gap is typically several electron-volts it is most unlikely that an electron will be thermally promoted from the valence band to the conduction band, and so an insulator remains insulating, even to very high temperatures (usually near to the

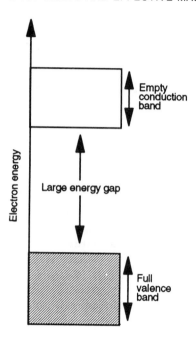

Figure 2.5   Energy-band diagram for an insulator.

material's melting point). An electric field is also unable to promote electrons across the band gap (refer to the semiconductor section, p. 14).

Will visible photons excite electrons across the energy gap in typical insulators? If we consider a 6 eV band-gap insulator (such as diamond), then the photon frequency would be 1.5×10^{15} Hz. This would correspond to ultraviolet radiation which is quite energetic. Ordinary visible photons would not carry enough energy (do the calculation).

From your understanding of semiconductor behaviour you will now see that an insulator is no more than a wide band-gap semiconductor. Some insulators can be doped with other materials just like semiconductors; diamond is such an example. Diamond can therefore behave like a very wide band-gap semiconductor which would operate at high temperatures, have low thermal noise, and good thermal conductivity, making it ideal for power devices. Silicon carbide is another example of an 'insulator' that can be doped to make semiconductor devices.

Finally, just as with a semiconductor, if enough energy is supplied to take an electron from the valence band to the conduction band, a hole will be available in the valence band for conduction as well as the electron in the conduction band (bipolar conduction).

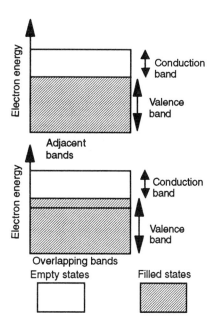

Figure 2.6 Energy-band diagrams for a metal.

2.4 Metals

Figure 2.6 shows an energy diagram or simple band diagram for a metal. There can be a little misunderstanding here since energy–momentum diagrams that we will come across a little later can also be called band diagrams, but we will use their full title of energy–momentum diagram or *E–k* diagram. Returning to Figure 2.6 we see that the band structure that typifies a metal is a valence band either just touching or overlapping the overlying empty conduction band. So in the case of the metal there are plenty of empty allowed energy-levels just above the valence band into which electrons can go if an electric field is applied. This is exactly what they do, and even small electric fields *can* provide enough energy to promote electrons into higher states where they are available for conduction. Considering semiconductors, insulators and metals, the metal is the odd one out because it has no energy gap between the valence and conduction bands. Both semiconductors and insulators have a forbidden energy gap between the valence and conduction bands, it is simply the magnitude of this gap that causes the difference in electrical characteristics between the two. There is one more point before we leave the metal. The electron promoted to the conduction band does not leave a positive charge carrier (hole) available for conduction in the valence band. If you look at the overall band diagram of the metal and don't split the two sections into a valence band and a conduction band, then you can see that what you have is a partly filled band, and partly filled bands conduct well. We are aware that conduction in most metals is by electrons but conduction mostly by holes does occur in some (divalent) metals such as beryllium, cadmium and zinc and this is due to their more complicated band structure. For the divalent metals there are a few empty states at the top of an otherwise filled band (as in Figure 2.4), and, as we saw for the semiconductor, this means that there are holes at the top of the band available for conduction. In some metals there may also be a few electrons in states above the hole states leading to bipolar conduction! These examples, although interesting, are now going a little further into band theory than I wish to cover.

2.5 The concept of effective mass

It cannot come as a surprise that an electron moving through a crystal does not behave in the same way as an electron travelling in a vacuum. If we apply an external force to an electron in a crystal, say by applying a voltage across the crystal, then the electron will accelerate, but at a different rate to a free electron experiencing the same electric field. This is *not* surprising; the electron in the crystal has to try and make its way through the crystal lattice, interacting with all the different potentials it would see depending on its direction through the lattice. What *is* surprising is that the complex interaction of the electron with the lattice can be taken into account, and we can use free-electron-type equations if we simply *change* the mass of the electron from its actual (vacuum) mass. This altered mass is called the electron's *effective mass* in the crystal and it depends on the direction it is travelling in the crystal. Although we don't have a free hole mass to compare with a hole in a crystal, the hole mass is also dependent on the direction in which the hole is travelling within the crystal. Why do we need to know about effective masses? There are at least four good reasons.

1 Up to this point in the chapter we have only considered energy-band diagrams. To appreciate what the effective mass means we have to look at *energy–momentum diagrams*. These energy–momentum diagrams will later show us why bulk silicon is a poor producer of photons and why gallium arsenide produces photons with a high efficiency. Also, an understanding of energy–momentum diagrams is all that is required in order to understand the basic workings of the Gunn diode, a source of microwaves that is simply a piece of gallium arsenide!

2 In studying carrier concentrations in semiconductors we shall come across something called the Fermi-level. In intrinsic material the Fermi-level would want to sit in the middle of the energy gap (the intrinsic Fermi-level) if it wasn't for the difference in *effective mass* between an *electron* and a *hole*.

3 We shall see that the effective mass of a hole actually turns out to be a *negative* quantity!! An empty electron state with a negative effective mass turns out to

behave as if it had a positive effective mass and a positive charge, i.e. it behaves as a hole. The concept of an effective mass therefore leads us to an even better understanding of the hole concept.

4 The energy–momentum diagram for an electron in a crystal, like the ordinary energy band-diagram, also shows regions where an electron is not allowed. These are the forbidden energy regions and energy–momentum diagrams show how these come about not by looking at bringing together discrete atoms, but by seeing that certain propagation conditions cannot be supported. We therefore gain a further insight into the origin of forbidden and allowed energy-levels!

I hope that these points illustrate the importance of a reasonably good understanding of this section of the chapter.

Before we finally get under way I will need to introduce another result from quantum mechanics. Sorry. But why do I introduce these things in dribs and drabs if they are needed for an understanding of device physics? Well, I have learned (the hard way) that it is too much to hit the new student with a lot of the quantum mechanical concepts at an early stage in their study. Brain seizure sets in and that's the end of it, which is a shame because this subject has so much to offer.

Let's get the quantum mechanical bit over with first. Particles, such as electrons, under certain conditions can show wavelike behaviour. What does this mean? Well, certain phenomena, such as diffraction and interference, can only be satisfactorily described by us at the moment if we consider them to be due to waves (particle equations don't work). If a particle can show wavelike behaviour, then we would expect that the most characteristic parameter associated with waves, *the wavelength*, is also somehow associated with particles. It is. The wavelength associated with particles is called the de Broglie wavelength, λ, and it links particle behaviour to wave behaviour. If it seems that things are getting a bit out of hand at this point I should mention that the diffraction of electrons (a purely wavelike phenomenon) is used routinely in semiconductor and materials analysis in the Transmission Electron Microscope (TEM). In the TEM, electrons can be diffracted off single crystals producing a diffraction pattern on a fluorescent screen which can be analysed to give information about the material under investigation. So we can (and routinely do) diffract electrons off single crystal materials, but what has this got to do with effective masses and forbidden energy regions? Electrons (and holes) travelling through a regular crystal lattice are diffracted just as X-rays are by crystal lattices. However, the electrons and holes will only be strongly diffracted if they satisfy the diffraction condition, which is a function of their momentum. Looked at in a slightly different way certain electron momenta are *not supported* by the crystal lattice and this gives rise to the forbidden energy bands we saw earlier in this chapter. We need a couple of equations and some pictures to see what's going on here. You should already be familiar with the *Bragg diffraction condition* for the diffraction of X-rays from single crystals. The Bragg condition is

$$n\lambda = 2d \sin(\theta) \tag{2.3}$$

where

n = the order of the diffraction,
λ = the wavelength of the X-ray,
d = the atomic spacing (lattice parameter),
θ = the incident angle of the X-ray beam.

Let's consider wave propagation normal to a set of atomic planes so that we can neglect the angular dependence and instead of X-rays let's now consider electrons. The simpler (angle free) diffraction equation becomes

$$n\lambda = 2d \tag{2.4}$$

At the values of λ given in equation (2.4) standing waves will be set up between the atomic planes. Since the wavelength λ of the electron is directly related to its momentum, electrons with this momentum cannot propagate through the crystal lattice. The electron under these conditions is unable to penetrate the crystal lattice and its velocity is zero. These conditions account for the forbidden energy bands!

Next we need to link the energy of an electron to its momentum if we are to draw energy–momentum diagrams. If we had a *completely free electron* we already know how its energy is related to its momentum; it's given by

$$E = \frac{p^2}{2m} \tag{2.5}$$

We now need to write the momentum p in terms of the particle's de Broglie wavelength. Momentum is related to wavelength according to

$$p = \frac{h}{\lambda} \tag{2.6}$$

where, h = Planck's constant.

The propagation constant of a wave is denoted by k, where

$$k = \frac{2\pi}{\lambda} \tag{2.7}$$

So we can write the energy–momentum equation (2.5) in terms of k instead of p according to

$$E = \frac{\hbar^2 k^2}{2m} \tag{2.8}$$

where $\hbar = h/2\pi$.

So we have an equation relating the energy of a particle E with the propagation constant k, which, apart from a multiplicative constant, is the same thing as its momentum. So if we plot diagrams of E against k we are in effect plotting particle energy against particle momentum, hence E–k diagrams are in fact energy–momentum diagrams. Let's look at some to see what's going on.

Figure 2.7 is a plot of equation (2.8), which is the energy–momentum

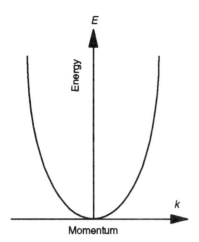

Figure 2.7 Energy–momentum diagram for a free electron.

relationship for a completely free electron. The E–k curve is parabolic and all values of energy (and momentum) are allowed. If we differentiate equation (2.8) twice with respect to k and rearrange we obtain

$$m = \frac{\hbar^2}{\mathrm{d}^2 E / \mathrm{d}k^2} \tag{2.9}$$

So we find that we can calculate the mass of a particle given its E–k diagram since the mass is just inversely proportional to *the curvature* of the E–k curve. Although we arrived at equation (2.9) by thinking about a totally free electron, the equation still holds good for electrons (or holes) in crystals, so we can be more general and write

$$m^* = \frac{\hbar^2}{\mathrm{d}^2 E / \mathrm{d}k^2} \tag{2.10}$$

where, m^* = the effective mass of the particle (electron or hole).

We now have sufficient background to start looking at some E–k diagrams that will introduce us to real semiconductor systems.

Figure 2.8 shows two hypothetical energy–momentum diagrams for an indirect and a direct band-gap semiconductor. We can gain a lot of information about the hypothetical semiconductor materials these diagrams depict from what we have just covered. Look at Figure 2.8(a) first, the indirect band-gap material. The first thing we note is that there is a maximum in the valence band curve at $k = 0$, and a minimum in the conduction band curve at a different k value. If an electron sitting near the conduction band minimum (where its energy is the lowest) is to recombine with a hole sitting near the valence band maximum, then the electron must lose some momentum (change its k value) in making this transition. In order for the electron to change its momentum another particle must be involved for energy and momentum conservation. This 'particle' is called a *phonon* and it is in fact just a lattice vibration. At any finite temperature the lattice atoms will be vibrating about their mean positions and these vibrations lead to the propagation of vibrational waves, phonons, throughout the crystal. The *momentum change* required for the electron to recombine with the hole in the indirect band-gap semiconductor comes from the *interaction* of the *electron* with a *phonon*.

Calculation

For GaAs, calculate a typical (band-gap) photon energy and momentum, and compare this with a typical phonon energy and momentum that might be expected with this material.

The band gap of GaAs is about 1.43 eV (so take this for the photon energy). Use equation (2.2) to estimate a typical photon wavelength:

Wavelength (microns) = 1.24/1.43 = 0.88 μm.

The photon momentum can be calculated from equation (2.6):

Momentum = $h/0.88 \times 10^{-6} = 7.53 \times 10^{-28}$ kg m s^{-1}

Now let's do the same sort of calculation for the phonon. We can use the same basic equations but we need to change a couple of values. Instead of the velocity of light, the phonons will travel at the velocity of *sound* in the material, approximately 5×10^3 m s^{-1}. Also, instead of the (long) photon wavelength the phonon wavelength is of the order of the material's lattice constant, which is 5.65×10^{-10} m in this case. From the equations you should see immediately that the phonon energy is going to be pretty small, because the velocity has dropped from the velocity of light to the velocity of sound, but the phonon momentum is going to be pretty big, because we are dividing by the lattice parameter, which is very small compared to the photon wavelength. Putting this altogether gives,

Phonon energy = 0.037 eV
Phonon momentum = 1.17×10^{-24} kg m s^{-1}

Compared with the photon result of,

Photon energy = 1.43 eV
Photon momentum = 7.53×10^{-28} kg m s^{-1}

We see that photons carry reasonable amounts of energy but very little momentum, hence *photon* transitions on the *E–k* diagram are *vertical* (no *k* change) transitions. On the other hand, phonons carry very little energy but have considerable momentum, hence the *indirect transitions* are facilitated by *phonon interactions*.

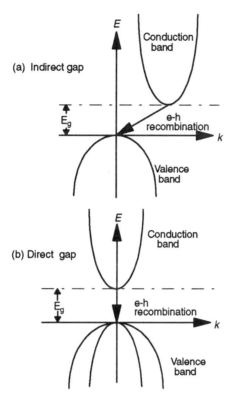

Figure 2.8 Energy–momentum diagrams for a semiconductor.

Clearly, the direct band-gap semiconductor does *not* require this three-particle interaction for electron–hole recombination since the recombination can take place at $k = 0$ with no momentum change. It is this difference that distinguishes the efficient photon producers, such as some of the 3–5 compound semiconductors (direct band-gap materials), from the highly inefficient photon producers such as bulk silicon and germanium (indirect band-gap materials). Unfortunately, an in-depth discussion of why this should be involves some heavy quantum mechanics but the basic reason is that transitions that are going to produce photons efficiently appear as *vertical* transitions on the *E–k* diagram because, as we have seen, photons themselves carry very little momentum. Such vertical transitions are clearly favoured by direct band-gap semiconductors. Electron–hole recombinations that involve phonons *without* the production of photons are called *radiationless transitions*. This type of transition is the most usual in the *elemental* semiconductors and their alloys. To make efficient light emitting diodes (LEDs) and LASERs the direct band-gap materials are the best (although LEDs *can* be made in indirect band-gap semiconductors that incorporate other materials).

We've already gleaned quite a lot of information from such a simple energy–momentum diagram but there's still plenty more to come!

From simple differential calculus theory you know that the curvature of a graph at a maximum point is a negative quantity and the curvature at a minimum point is a positive quantity. We have a minimum in the E–k curve in the *conduction band* and since the sign of the *effective mass* of a particle can be obtained directly from the sign of the curvature of the E–k curve (see equation (2.10)), we immediately see that the particles (electrons) sitting near the minimum have a *positive effective mass*. Now look at the valence band maximum. Clearly, particles sitting at the valence band maximum (holes) have a *negative effective mass* by exactly the same argument. A negative mass implies that a particle will go *'the wrong way'* when an external force is applied and this is what the holes do; they go in the opposite direction to the electrons in an electric field!

We haven't finished yet. I have deliberately drawn a greater curvature for the conduction band of the indirect band-gap semiconductor than for the valence band in Figure 2.8(a). This implies that the effective-mass of the electron is *smaller* in magnitude than the hole effective mass in this case. But what is going on in the *valence band* in Figure 2.8(b)? There are two curves of different curvature. This is typical of compound semiconductors such as GaAs and the two curves correspond to the *light-hole* band and the *heavy-hole* band, named as such for obvious reasons. Which is which?

As a final example of an energy–momentum diagram consider Figure 2.9. Remember that a semiconductor's band gap is the smallest energy separation between the valence and conduction band edges, and now look at Figure 2.9 again. In Figure 2.9(a) the smallest energy gap is between the top of the valence band at $k = 0$ and one of the conduction band minima away from $k = 0$. Hence, the semiconductor shown in Figure 2.9(a) is an *indirect band-gap* semiconductor. Now look at Figure 2.9(b). The smallest energy difference lies between the conduction band minimum and the valence band maximum, both of which occur at $k = 0$. This is a *direct band-gap* semiconductor. This final example is *not* purely hypothetical. The compound semiconductor aluminium gallium arsenide (AlGaAs) behaves like this. For small aluminium concentrations of a few percent, AlGaAs is a direct band-gap material. However, we would like to push up the aluminium concentration because as we do so the conduction band minimum increases in energy and we get a bigger band-gap material. If we were making an LED or a LASER this means we would get higher-energy (shorter-wavelength) photons out of the device. But, as Figure 2.9 shows, we can't just keep adding aluminium to get shorter and shorter wavelength devices; eventually the conduction band minimum at $k = 0$ rises above the conduction band minima lying away from $k = 0$, and the semiconductor *goes indirect gap*. When this happens the photon production efficiency falls off drastically and we no longer have a photon emitting device.

As a last observation of Figure 2.9, notice that the two conduction bands (often called satellite valleys) on either side of the conduction band centred at $k = 0$,

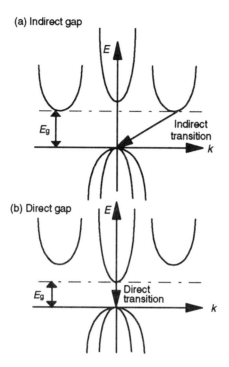

Figure 2.9   AlGaAs $E-k$ diagrams.

have a *smaller curvature* than the central conduction band. From equation (2.10) this means that the electron effective mass in these satellite valleys is *greater* than the electron effective mass in the central (valley) conduction band. The higher effective-mass electrons in the satellite valleys turn out to be *less mobile* than the 'lighter' electrons in the central valley. This fact has far reaching consequences that we will consider in Chapter 5.

CHAPTER THREE

Carrier concentrations in semi-conductors

3.1 Donors and acceptors

Recall from Chapter 2 that for *intrinsic* semiconductors, that is, semiconductors without any other different atoms added, conduction occurred via electrons in the conduction band and holes in the valence band. For intrinsic material the electrons and holes came in equal numbers, since one conduction electron gave rise to one valence band hole, and we had *bipolar* (two-carrier) conduction, with the total current being carried by *equal* numbers of electrons and holes (*intrinsic conduction*).

Hence, for intrinsic semiconductors:

$$n = p = n_i \qquad (3.1)$$

where

    n = concentration of electrons per unit volume,
    p = concentration of holes per unit volume,
    n_i = the intrinsic carrier concentration of the semiconductor under consideration.

The intrinsic carrier concentration n_i depends on the semiconductor material and the temperature. For silicon at $300\,\mathrm{K}$, n_i has a value of $1.4 \times 10^{10}\,\mathrm{cm}^{-3}$. Clearly, equation (3.1) can also be written as

$$n \cdot p = n_i^2 \qquad (3.2)$$

The importance of *this* equation is that it is valid for *extrinsic* as well as *intrinsic* material. I am concentrating on this because in real devices, although intrinsic material is occasionally used, more often than not we want material that carries current mostly using either electrons *or* holes, but not both! How can we ensure the predominance of one carrier type in a semiconductor? In order to do this we need to introduce other atoms into the semiconductor lattice, turning it from *intrinsic* (undoped material) to *extrinsic* (doped) material. The addition of these other atoms is called doping, and these other atoms are called dopants or impurities, centres, impurity levels, deep levels or shallow levels and a host of other names. Whatever they are called it's all the same thing, the addition of different atoms to modify the conductivity of the intrinsic semiconductor. At this point it's worth mentioning that 'impurities' is a bit of a misnomer since they are being deliberately added to modify conductivity, but impurities is a label you will commonly see.

What atoms can we add to a semiconductor to modify its conductivity? We need to go back to some Chapter 1 material in order to see the answer to this. First of all let us look at the intrinsic silicon case in Figure 3.1. I'm not showing any crystal structure in this diagram; it's a flat two-dimensional representation, but what it does show is the bonding arrangement of the four-valent silicon atoms. Silicon, like its neighbours in group 4 of the periodic table (carbon and germanium) is a covalently (electron sharing) bonded solid and shows all the characteristic physical properties of covalent solids. We are now in a position to see the effect of introducing group 3 *acceptor centre* and group 5 *donor centre* atoms into the silicon lattice. Figure 3.2 shows the effect of adding a group 3 atom (boron in this case) to the silicon lattice. The boron has gone onto a silicon site but it only has three valence electrons, one short for completing the full outer-shell, rare-gas configuration. This deficiency of one electron per impurity atom is in fact the same thing as introducing one hole per impurity atom! This vacant electron state can accept a nearby valence electron (hence the term 'acceptor') and in this manner the vacant electron state appears to move, see Figure 3.3. There are therefore *two ways* of introducing holes into semiconductors. One way is to promote an electron from the valence band to the conduction band (by thermal or electromagnetic energy), which leaves a hole in the valence band. This process produces equal numbers of electrons and holes. The other way is to introduce electron deficient (with regard to the silicon bonding requirements) atoms into the silicon lattice. Adding these acceptors to the silicon lattice also introduces holes into the valence band, but we have to look a little further to see how this occurs. We need a physical picture to appreciate what is happening with these acceptor states. Let us think a little about these vacant electron states. Very little energy is required to move a nearby *valence* electron into the vacant state. We can represent this situation as shown in the energy diagram of Figure 3.4. Here the vacant state (acceptor level) lies just above the *valence band*, therefore just a *little* energy can move a valence electron into the vacant state, as required. However, having lost the electron from the valence band to the acceptor level we are now left with a hole in the valence band which, as we saw in Chapter 2, is available for conduction. Let's go even further. The acceptor level is so close to the valence band edge that very little energy is required to produce the holes; look at the diagram, it's very much less than the band-gap energy. The point is that there is enough thermal energy *even* at room-temperature to move electrons into the acceptor level so that *each acceptor* atom that we introduce gives rise to a *hole* immediately. In this discussion we are referring to *shallow* acceptors, those that lie very close to the valence band edge. Acceptors that introduce energy-levels nearer to the middle of the band gap are called *deep* acceptors, or *deep levels*. These are very important for other purposes, as we shall see later, but because they lie further away from the valence band edge, room-temperature may not supply enough thermal energy to move electrons from the valence band to *all* of the introduced deep acceptors. This partial occupancy of the deep-level acceptors means that not every acceptor gives rise to a conduction hole in the valence band, which may be important if you are trying to modify a semiconductor's conductivity using deep levels. Returning to the simpler, shallow acceptor-level case, note one other very important fact. We have introduced holes into the valence band, available for conduction, and we have done so without putting any extra electrons in the conduction band! We have produced *p-type silicon* that conducts *mostly* by holes. There will always be a few electrons around but their numbers may be many orders of magnitude less than the holes. We have achieved our aim; we have produced semiconductor material that uses holes for conduction without adding an equal number of electrons. There is one last thing to mention before we look at producing material that conducts mostly by electrons, *n-type material*. The acceptor *does not* partake in conduction itself! Once it has received its electron and attained its rare-gas configuration it is electrically 'inert'. This is not necessarily true for *very high* concentrations of acceptors. In this case you can get what is called *impurity band conduction* and *hopping* occurring, where the level *is* actually involved in conduction, but an understanding of this is going beyond an introductory-level text.

It only remains to look at the case of adding electrons to semiconductors without adding additional holes to make n-type material. I chose to introduce you

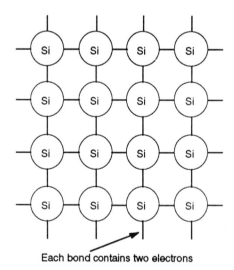

Each bond contains two electrons

Figure 3.1   Bonding in intrinsic silicon.

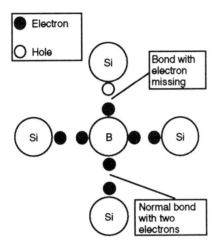

Figure 3.2   Boron bonding in silicon.

to n-type doping after p-type doping because conceptually it's much easier to see what's going on. The difficult p-type case has been dealt with first!

Figure 3.5 shows the bonding configuration that results when a group 5 impurity atom is added to the silicon lattice and sits on a lattice site. I keep mentioning the impurity atom sitting *on* a silicon site because it is possible for impurities to go into silicon without sitting on a silicon site. These *interstitial* impurities are more difficult to analyze and introduce different energy-levels into the semiconductor's band gap. From Figure 3.5 you can see that we have one electron too many to satisfy the covalent bonding requirements and this extra electron is in fact very weakly bound to the impurity, which we are taking as phosphorus in this case (it could have been arsenic if we had wished). In a manner exactly analogous to the p-type case we can draw a suitable energy-level diagram, as seen in Figure 3.6. Here the impurity atoms form a level just *below* the conduction band edge; an electron therefore needs very little energy to move from the impurity centre up into the conduction band where it is available for conduction. The impurity centre is said to *donate* an electron to the conduction band and these are called *donor centres*. Note that the loss of an electron from the donor centre *does not* leave a hole at the donor!! The donor, together with its neighbouring silicon atoms has its full complement of electrons (to attain a rare-gas configuration) *without* the extra electron. The extra electron was surplus to bonding requirements; it has been readily given up to the conduction band and is now available for conduction. As with the acceptor doping case, the donor level *does not* itself partake in electrical conduction unless the concentration of donors comes to within a few percent of the silicon atom concentrations (i.e. VERY big!). We have now introduced electrons into the conduction band of silicon, producing n-type material, without adding an equal number of holes to the

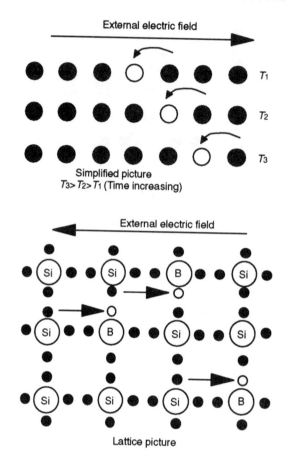

Figure 3.3 Hole movement due to an electric field.

valence band. We are thus now able to make silicon (and some other semiconductors) conduct mainly by electrons (n-type) or mainly by holes (p-type) if we want. We shall see why we want to do this when we look at real devices. One reason we might want to dope the silicon is obvious right now. Recall that intrinsic silicon can only conduct by the thermally (or optically) generated electrons and holes. At room-temperature these numbers are pretty small for pure silicon and so intrinsic silicon, as a pure conductor, conducts very badly under these circumstances. Adding donors or acceptors clearly adds a number of electrons (or holes) in proportion to the number of centres added and so the *extrinsic* (or doped) silicon can be made quite conducting and, therefore, more useful for incorporation in electronic devices.

Let's recap a couple of important points that I have 'glossed' over without the mention they deserve.

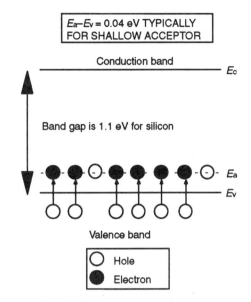

$E_a - E_v = 0.04$ eV TYPICALLY
FOR SHALLOW ACCEPTOR

Conduction band
E_c

Band gap is 1.1 eV for silicon

E_a
E_v

Valence band

○ Hole
● Electron

Figure 3.4 Shallow acceptor in silicon.

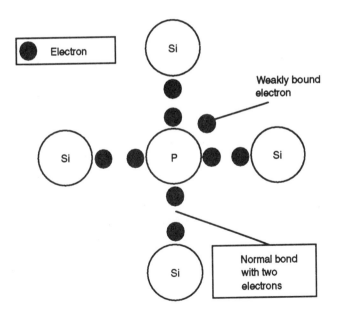

● Electron

Si

Weakly bound
electron

Si   P   Si

Si

Normal bond
with two
electrons

Figure 3.5 Phosphorus bonding in silicon.

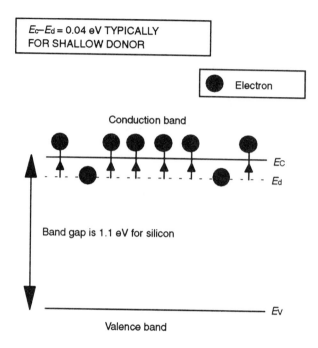

Figure 3.6   Shallow donor in silicon.

1   The donors (and acceptors) have introduced allowed energy-levels *in the forbidden energy-gap!* The energy gap is forbidden *only* for *pure* material, i.e. *intrinsic* silicon. Once impurity atoms are added to the silicon the perfect crystal lattice is disrupted by these additions (the potential energies change) and it is no longer forbidden to have allowed states (energy-levels) in the once forbidden energy gap.

2   p-type or n-type silicon is *electrically neutral*. Think about it. A neutral atom has gone into neutral silicon. Conduction may be altered by the added weakly bound electrons or holes, but *overall* there is *charge neutrality. Locally* there may *not* be charge neutrality; when the donor centre loses its electron (becomes ionized) it no longer has its full complement of electrons for charge neutrality. The donor centre when ionized is positively charged. When it has its carrier on it (non-ionized) the donor is neutral. This is in fact the definition of a donor centre. We have the converse situation for the acceptor centre. When the acceptor is not ionized (still has its hole) it must be neutral because that's how it went into the lattice. However, if we ionize the acceptor, which means we remove the hole (or, equivalently, it grabs an electron from the valence band), it becomes *negatively* charged. Look at Figure 3.7 to see this pictorially. This may be taken as the definition of an acceptor level. These points will be made more clear, and their importance seen, when we look at p–n junction diodes where we have

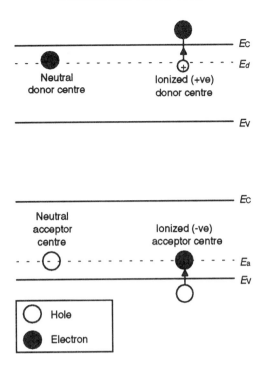

Figure 3.7 Donor and acceptor charge states.

ionized donor and acceptor centres in the diode's *depletion region*. We must therefore have some *local charge* in the depletion region of the diode and we shall see how this leads to a potential barrier and rectifying action.

3 We are not restricted in adding group 3 or group 5 elements to silicon. We could add any element we please. Clearly, some will be electrically active (introduce energy-levels into the energy gap), and some may not. For example, oxygen in silicon can produce donor levels and acceptor levels. Germanium and carbon, being in the same group as silicon tend to alloy with silicon producing a semiconductor with a different energy gap. Silicon carbide has a bigger energy gap than silicon and can be used in high-temperature applications (it can be doped p- and n-type and transistors can be made of it). Silicon–germanium has a smaller energy gap than silicon and is used as the *base region* in a silicon transistor called a *heterojunction bipolar transistor*.

Finally, in this section, we need to introduce some new symbols to discuss carrier concentration in semiconductors. In p-type semiconductors there are far more holes than electrons. Let's calculate how many more.

Calculation

Calculate the hole and electron densities in a piece of p-type silicon that has been doped with 5×10^{16} acceptor atoms per cm^3.

If the acceptor impurity is a shallow centre then each atom will give one hole to the silicon, hence the hole concentration is $p = 5 \times 10^{16} cm^{-3}$.

For the electron concentration return to equation (3.2) which I stated (without proof) was OK for extrinsic material. Then the electron concentration n is given by

$$n = n_i^2/p = 3.9 \times 10^3 \text{ electrons per } cm^3$$

Note that this electron concentration lies well below the intrinsic electron concentration for p-type material. This imbalance in carrier densities is usual in extrinsic material. The more holes you put in the less electrons you have and vice-versa. The balance has to be maintained by equation (3.2).

So we see that in p-type material there are many more holes than electrons, the holes are the *majority carriers* and the electrons are the *minority carriers*. In n-type material electrons are the *majority carriers* and holes are the *minority carriers*. This leads to the following four possibilities:

(a) Electron concentrations in n-type material which are denoted n_n.
(b) Electron concentrations in p-type material which are denoted n_p.
(c) Hole concentrations in p-type material which are denoted p_p.
(d) Hole concentrations in n-type material which are denoted p_n.

3.2 Fermi-level

This section presents the Fermi–Dirac equation and discusses the Fermi-level for the first time. Unfortunately, we need to expend quite a lot of intellectual energy, specifically in trying to come to grips with the idea of the Fermi-level, with relatively small reward for our efforts. In fact, there are only three results that we shall use elsewhere resulting from this exercise:

1 We shall see that the equation $np = n_i^2$ is valid for extrinsic as well as intrinsic semiconductors. Earlier, this has been assumed without proof.
2 We shall see how the Fermi-level moves towards the conduction band edge with increasing donor concentration, or towards the valence band edge with increasing acceptor concentration. This fact will be used mainly in the construction of device energy-band diagrams.
3 Finally an important result that we shall find is that the Fermi-level moves towards the middle of the band gap as the temperature increases. This is where the Fermi-level wants to be for *intrinsic* material so it means that as the temperature is increased, no matter whether the semiconductor is p- or n-type,

it tends to go intrinsic at high temperature. Since intrinsic material is neither p-nor n-type it means that we have lost the extrinsic characteristic of the material and any device that relies on discrete p- and n-type regions in its structure will fail at high temperatures.

Those are the main reasons for this section, let's start with a definition of the Fermi-level and go on from there. The Fermi-level is simply a *reference* energy-level at which the probability of occupation by an electron is one-half (0.5). As the Fermi-level is only a reference level it can appear at different places on a semiconductor energy-band diagram depending on the doping conditions. We can gain a little more insight into the meaning of Fermi-level from the Fermi–Dirac function, which gives us the probability of an energy-level at energy E being occupied by an electron (obviously there has to be an allowed level at E in the first place if an electron can exist there!). In full, the Fermi–Dirac function is

$$P_{fd}(E) = \frac{1}{1 + \exp\dfrac{(E - E_F)}{kT}} \tag{3.3}$$

Let's spend just a little time looking at this equation. In words it says that the probability of occupation by an electron of an energy-level at energy E is an inverse exponential function involving the difference in energy between the energy-level under consideration, E, and the *Fermi-level*, E_F. If we consider the energy-level E to be *at* the Fermi-level then we obtain

$$P_{fd}(E_F) = \frac{1}{1 + \exp(0)} = \frac{1}{2} \tag{3.4}$$

so that, as we defined earlier, the probability of finding an electron *at* the Fermi-level is one-half. If $P_{fd}(E)$ is the probability of finding an electron at energy E, then $1 - P_{fd}(E)$ must be the probability of *not* finding an electron at energy E, or the probability of finding *a hole* at energy E.

Calculation

Show that for a small incremental energy *above* the Fermi-level the probability of occupation by an electron is the same as for the occupation by a hole for the same small incremental energy difference *below* the Fermi-level in a semiconductor. For clarification on the physical situation corresponding to the above question, refer to Figure 3.8.

The probability that an electron can exist at an energy dE above the Fermi-level is given by

$$P_{fd}(E_F + dE) = \frac{1}{1 + \exp\left(\dfrac{dE}{kT}\right)}$$

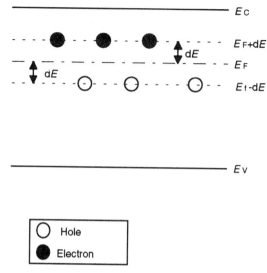

Figure 3.8   Diagram for calculation.

and the probability that a hole can exist at an energy dE *below* the Fermi-level is given by

$$1 - P_{fd}(E_F - dE) = 1 - \frac{1}{1 + \exp - \left(\dfrac{dE}{kT}\right)}$$

which, on carrying out the subtraction, gives

$$1 - P_{fd}(E_F - dE) = \frac{1}{1 + \exp\left(\dfrac{dE}{kT}\right)}$$

in agreement with the electron occupation probability $P_{fd}(E_F + dE)$

The above calculation indicates the symmetry of the Fermi–Dirac function about the Fermi-level. Let's finally look at the temperature dependence of the function starting with its behaviour at $T = 0\,\mathrm{K}$. If we put $T = 0\,\mathrm{K}$ in the Fermi–Dirac function we get either

$$P_{fd}(E) = \frac{1}{1 + \infty} = 0 \tag{3.5}$$

if $E > E_F$ or

$$P_{fd}(E) = \frac{1}{1 + 0} = 1 \tag{3.6}$$

if $E < E_F$.

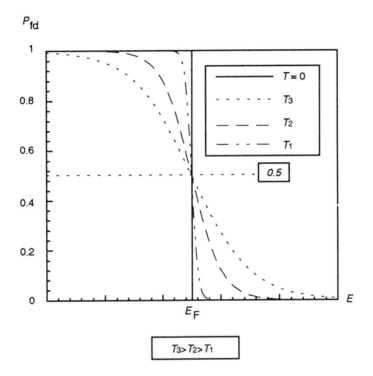

Figure 3.9 Fermi–Dirac function.

So at $0\,\mathrm{K}$ all levels below the Fermi-level are filled, and all levels above the Fermi-level are empty. Raising the temperature causes some levels above E_F to become filled and, therefore, since the function is symmetric, some levels below E_F to empty. A plot of probability as a function of energy for the Fermi–Dirac function is shown in Figure 3.9 for several different temperatures.

We've looked long enough at the Fermi–Dirac equation and the Fermi-level to have a basic understanding of what's going on. It's now time to move on to the equations that describe carrier concentrations in semiconductors.

3.3 Carrier concentration equations

We shall start by going straight into the relevant equations, without derivation, so that we can get on with the practicalities of the subject. A fuller explanation into the origin of the equations is given in Appendix 1. The equations describing the *number density* of electrons in the conduction band (the electrons available for conduction) are as follows:

$$n = 2\left(\frac{2\pi m_n^* kT}{h^2}\right)^{\frac{3}{2}} \exp-\left(\frac{E_C - E_F}{kT}\right) \tag{3.7}$$

$$n = N_C \exp-\left(\frac{E_C - E_F}{kT}\right) \tag{3.8}$$

$$n = n_i \exp\left(\frac{E_F - E_i}{kT}\right) \tag{3.9}$$

where, m_n^* is the electron effective mass and E_C is the conduction band edge energy.

The corresponding equations for the density of holes in the valence band are

$$p = 2\left(\frac{2\pi m_p^* kT}{h^2}\right)^{\frac{3}{2}} \exp-\frac{E_F - E_V}{kT} \tag{3.10}$$

$$p = N_V \exp-\left(\frac{E_F - E_V}{kT}\right) \tag{3.11}$$

$$p = n_i \exp\left(\frac{E_i - E_F}{kT}\right) \tag{3.12}$$

where, m_p^* is the hole effective mass and E_V is the valence band edge energy.

In the above equations, N_C is called the *effective density of states* in the conduction band and N_V is called the *effective density of states* in the valence band. These are *not* the *actual* density of states (number density of allowed energy-levels) in the band, but they are a modification of this; see Appendix 1. The energy E_i is the *intrinsic Fermi-level*. It is the position that the Fermi-level occupies in an intrinsic (undoped) semiconductor. The only way to come to grips with what these equations are telling us is by a series of explanatory calculations.

Calculation

Calculate the position of the Fermi-level in an *intrinsic* semiconductor, i.e. calculate the position of E_i.

If we put $n = n_i$ in equation (3.7) then, by definition, E_F becomes E_i. Similarly, in equation (3.10), change p to p_i and E_F to E_i. Then, since $n_i = p_i$ for intrinsic material:

$$2\left(\frac{2\pi m_n^* kT}{h^2}\right)^{\frac{3}{2}} \exp-\left(\frac{E_C - E_i}{kT}\right) = 2\left(\frac{2\pi m_p^* kT}{h^2}\right)^{\frac{3}{2}} \exp-\left(\frac{E_i - E_V}{kT}\right)$$

Rearranging the above equation gives us

$$\left(\frac{m_n^*}{m_p^*}\right)^{\frac{3}{2}} = \exp\left(\frac{E_V + E_C - 2E_i}{kT}\right)$$

Hence,

$$E_V + E_C - 2E_i = \frac{3}{2} kT \ln\left(\frac{m_n^*}{m_n^*}\right)$$

Finally, making E_i the subject of the equation,

$$E_i = \frac{E_V + E_C}{2} + \frac{3}{4}kT \ln\left(\frac{m_p^*}{m_n^*}\right)$$

Figure 3.10 illustrates the meaning of this equation for E_i. There are some interesting points to be gleaned from the equation (and the diagram). Since, in general, the electron and hole effective masses are *unequal*, the intrinsic Fermi-level *does not* lie in the middle of the band gap. This is one fact stated earlier that we wished to prove. The intrinsic Fermi-level *would* lie in the middle of the band gap if the part of the equation containing the effective masses disappeared. This will occur at $T = 0\,\text{K}$, and for equal effective-masses. Finally, convince yourself by looking at Figure 3.10 that $(E_V + E_C)/2$ *does* lie in the middle of the energy gap.

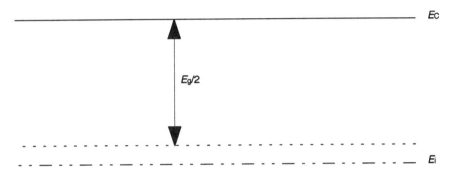

Figure 3.10 Intrinsic Fermi-level.

Calculation

Show that the equation $np = n_i^2$ holds for *extrinsic* as well as *intrinsic* semiconductors.

Simplistically, we could choose equations (3.9) and (3.12) and show

$$np = n_i^2 \cdot \exp\left(\frac{E_i - E_F}{kT}\right) \cdot \exp\left(\frac{E_F - E_i}{kT}\right) = n_i^2$$

This is not particularly enlightening so let's go back one step and use equations (3.8) and (3.11) instead. Now,

$$np = N_C N_V \cdot \exp -\left(\frac{E_C - E_F}{kT}\right) \cdot \exp -\left(\frac{E_F - E_V}{kT}\right)$$

And multiplying out the exponential terms gives

$$np = N_C N_V \cdot \exp -\left(\frac{E_C - E_V}{kT}\right)$$

The first thing to notice is that E_F has disappeared, which is useful since if we want to find a constant parameter (n_i^2) we can't have a quantity (E_F) that varies with doping concentrations. Next, notice that $E_C - E_V = E_g$, the energy gap, so we can rewrite the above equation as

$$np = N_C N_V \cdot \exp -\frac{E_g}{kT}$$

And since we know that this must equal n_i^2, then,

$$n_i = \sqrt{N_C N_V} \cdot \exp -\frac{E_g}{2kT}$$

giving us an intrinsic carrier concentration which only depends on the temperature and the type of semiconductor under consideration.

Finally, put the numbers into this equation for n_i using $m_n^* = 1.18$, $m_p^* = 0.59$ and $T = 300\,\mathrm{K}$. You will of course have to change N_V and N_C into their basic components as in (3.7) and (3.10)!! Recall that we used an n_i of the order of $10^{10}\,\mathrm{cm}^{-3}$ earlier. How does your calculation compare to this? (You should get something around $8 \times 10^9\,\mathrm{cm}^{-3}$.)

Worked example

How does the position of the Fermi-level change with

(a) increasing donor concentration,
(b) increasing acceptor concentration?

Starting with part (a) first we shall use equation (3.8). If n is increasing then the quantity $E_C - E_F$ must be decreasing, i.e. as the donor concentration goes up the Fermi-level *moves towards* the conduction band edge E_C, as shown in Figure 3.11. This also introduces a very important point. Unfortunately equations (3.7)–(3.12) aren't valid for all doping concentrations! As the Fermi-level comes to within about $3kT$ of either band edge the equations are no longer valid because they were derived by assuming the simpler Maxwell–Boltzmann statistics rather than the proper Fermi–Dirac statistics. The reason for doing this is covered in Appendix 1. With this proviso in mind let's look at part (b) of the question.

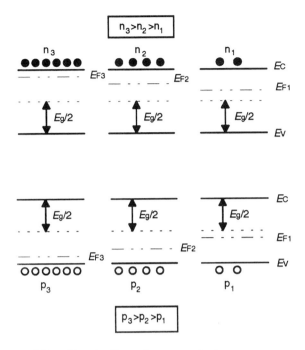

Figure 3.11 Change of Fermi-level with carrier concentration.

Equation (3.11) tells us that as the acceptor concentration increases, the Fermi-level moves towards the *valence band edge*. These results will be used in our construction of device (energy) band diagrams.

Thought experiment

We shall find out what happens to the Fermi-level as the temperature increases in a non-mathematical way by thinking about the physics of the problem. As we increase the temperature of a p- or n-type doped semiconductor the number of electron–hole pairs increases because more thermal energy is available to promote an electron from the valence band to the conduction band. So as we increase the temperature there will come a point where the number of thermally generated electron–hole pairs exceeds the number of deliberately introduced dopant carriers. At this point the semiconductor starts to lose its characteristic doping type, and as the electron–hole pairs greatly exceed the doping, then the semiconductor's doping-type is lost altogether. At a temperature where the electron–hole pairs greatly outnumber the doping carriers the material will look *intrinsic* because its properties are like those of an intrinsic semiconductor, since the added dopant carriers are insignificant compared to the natural thermally generated electron–hole pairs. Since the material is behaving intrinsically, the

Fermi-level is at E_i, near the middle of the band gap. So in p-type material, as the temperature increases, the Fermi-level will move away from a position near the valence band edge, towards the middle of the energy gap. For n-type material the Fermi-level moves away from a position near the conduction band edge, towards the middle of the gap, with increasing temperature.

Calculation

This last calculation of the chapter will tie together most of the facts we have gained from equations (3.7)–(3.12).

A piece of silicon is doped with 5×10^{16} Boron atoms per cm^3. Calculate the electron concentration at 300 K. What is the position of the Fermi-level with respect to the valence band edge and the intrinsic Fermi-level?

You will see this *type* of question in many texts and exam papers for physical electronics. The key starting point is that you must consider that *all* the introduced centres are ionized (electrically active), so that in this case we have 5×10^{16} holes per cm^3 in the piece of silicon (remember Boron is an acceptor centre).

We now use $np = n_i^2$ to calculate the electron concentration as

$$n = 2000\,\text{cm}^{-3}$$

by taking n_i^2 as 10^{20} per cm^3.

Assuming that N_V is $10^{19}\,\text{cm}^{-3}$, and that kT at 300 K is 25.8 meV, then

$$kT \cdot \ln\frac{p}{N_V} = -(E_F - E_V) = -0.137\,\text{eV}$$

so that the Fermi-level lies above the valence band edge by 0.137 eV.

The equation for calculating the difference between the intrinsic Fermi-level and the Fermi-level is given by

$$kT \cdot \ln\frac{p}{n_i} = E_i - E_F = 0.398\,\text{eV}$$

so that the intrinsic Fermi-level lies above the Fermi-level by 0.398 eV. The situation is shown in Figure 3.12. By adding these two results together we get the difference between the intrinsic Fermi-level and the valence band edge as 0.535 eV, which isn't quite in the middle of silicon's energy gap (0.55 eV at 300 K).

3.4 Donors and acceptors both present

It seems most unlikely that, in a piece of semiconductor, only donors or only acceptors will be present. Trace impurities incorporated during the growth of semiconductor crystals ensures that in general *both* donors and acceptors are

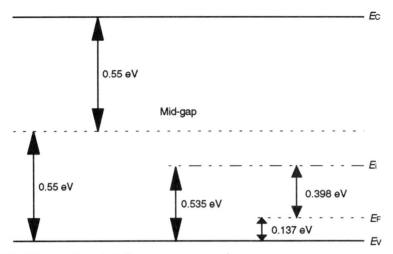

Figure 3.12 Diagram for calculation.

present, although one will outnumber the other. How do we handle this? Actually, it's a lot simpler than you might imagine. If we have a piece of silicon in which the shallow donor concentration N_D is significantly greater than the shallow acceptor concentration N_A, then the material appears to be *n-type* according to

$$n_n = N_D - N_A \qquad (3.13)$$

Similarly if a piece of silicon has significantly more shallow acceptors than shallow donors,

$$p_p = N_A - N_D \qquad (3.14)$$

and the material will appear to be *p-type*.

From the above it should be clear that n-type material can be converted to p-type and vice-versa, by the addition of excess dopant atoms of the appropriate type. This is in fact how p–n junction diodes are actually fabricated. A piece of silicon is locally *type-converted* by diffusing, or ion implanting, the required amount of dopant into material of the opposite type.

So for material that contains *both* donors and acceptors, assuming complete ionization, and overall charge neutrality we can write

$$p + N_D^+ = n + N_A^- \qquad (3.15)$$

which says that the total positive charge density, due to free holes and ionized donor centres, equals the total negative charge density, due to free electrons and ionized acceptor centres. Also from equation (3.2) we know that $np = n_i^2$ and we can therefore solve (3.15) and (3.2) simultaneously to give

$$n_n = \tfrac{1}{2}(N_D - N_A + [(N_D - N_A + [(N_D - N_A)^2 + 4n_i^2]^{\frac{1}{2}}) \qquad (3.16)$$

and

$$p_n = \frac{n_i^2}{n_n} \qquad (3.17)$$

where (3.16) and (3.17) are for material with $N_D > N_A$, i.e. for n-type material.

We can go through exactly the same exercise for the case where $N_A > N_D$, i.e. for p-type material and we would obtain,

$$p_p = \tfrac{1}{2}(N_A - N_D + [(N_A - N_D)^2 + 4n_i^2]^{\frac{1}{2}}) \qquad (3.18)$$

with the minority carrier concentration being given by

$$n_p = \frac{n_i^2}{p_p} \qquad (3.19)$$

Using the above set of equations you could calculate the Fermi-level position as a function of temperature for different doping concentrations. From our earlier thought experiment you might guess that the temperature at which material goes intrinsic is dependent on the doping concentration. Furthermore, the higher the doping concentration, the higher the temperature before intrinsic behaviour sets in.

That series of calculations completes our initial work on carrier concentrations in semiconductors. Now that we know how many free carriers there are per unit volume in a semiconductor, we can start to calculate how much current is flowing, and therefore how devices will behave. However, currents may flow by different mechanisms, namely drift and diffusion. Before we see how devices function in detail we need to know how drift and diffusion currents flow in semiconductors.

CHAPTER FOUR

Conduction in semiconductors

From the work we covered in the last chapter we now have some idea of the number density of charge carriers present in semiconductor materials. Since current is the rate of flow of charge, we shall be able to calculate currents flowing in real devices once we get to grips with the two mechanisms that cause charges to move in semiconductors. The two mechanisms we shall study in this chapter are *drift* and *diffusion*.

4.1 Carrier drift

Current carriers (i.e. electrons and holes) will move under the influence of an electric field because the field will exert a force on the carriers according to

$$\mathbf{F} = q\mathbf{E} \tag{4.1}$$

In this equation you must be very careful to keep the signs of $\mathbf{E}$ and q correct. Since the carriers are subjected to a force in an electric field they will tend to move, and this is the basis of the drift current. Figure 4.1 illustrates the drift of electrons and holes in a semiconductor that has an electric field applied. In this case the source of the electric field is a battery. Note the *direction* of the current flow. The direction of current flow is defined for *positive* charge carriers because long ago, when they were setting up the signs (and directions) of the current carriers they had a 50:50 chance of getting it right. They got it wrong, and we *still* use the old convention! Current, as you know, is the rate of flow of charge and the equation you will have used should look like

$$I_d = \mathrm{n}Av_dq \tag{4.2}$$

where

I_d = (drift) current (amperes),
n = carrier density (per unit volume),
A = conductor (or semiconductor) area,
v_d = (drift) velocity of carriers,
q = carrier charge (coulombs),

which is perfectly valid. However, since we are considering charge movement in an electric field it would be useful to somehow introduce the electric field into the equations. We can do this via a quantity called the mobility, where

$$v_d = \mu E \tag{4.3}$$

In this equation:

μ = the drift mobility (often just called the mobility) with units of $cm^2 V^{-1} s^{-1}$,
E = the electric field ($V\,cm^{-1}$).

You should be used to the mixture of SI and CGS units by now. It is unfortunate that cm units still persist in semiconductor physics but you will come across this very often. Note the units of mobility. They can cause confusion. The easiest way to remember the units of mobility is to remember equation (4.3). Mobility is clearly a measure of how easily charge carriers move (how mobile they are) under the influence of an electric field. This fundamentally important parameter is also directly related to the diffusion equations via an equation called the *Einstein relation*. This being the case we will need to look at carrier mobility in some detail.

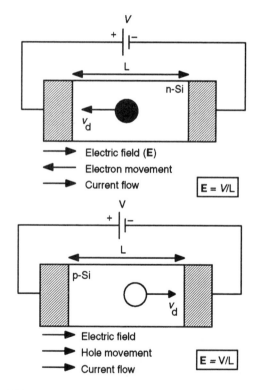

Figure 4.1 Carrier drift.

4.2 Carrier mobility

In equation (4.3) we have a *macroscopic* understanding of carrier mobility. However, if we want a fundamental understanding, we need to know what is going on at the *microscopic* level, i.e. we need to know what the carriers themselves are doing.

We'll kick off with a very interesting fact. If the electrons (or holes) were moving through a *perfect* crystal lattice with the lattice atoms stationary, then nothing would impede the carrier's progress and the crystal would appear to have zero resistance to current flow, like a superconductor. The reasoning behind this comes from some quantum mechanics that we unfortunately cannot cover, but it shows that the perfect periodic potential field seen by a current carrier, in a perfect crystal lattice, does not impede carrier movement at all. It is only *interruptions* in the perfect periodicity in the potential field that upsets things. Our *real* semiconductor crystal is going to look anything but perfect. A real piece of silicon will contain many *impurities* sitting in the lattice, and since the impurity is a different atom it will have a different potential field associated with it, and this will upset the perfect periodicity we require for zero resistance. Also, there will be lattice sites where the silicon atom is actually missing, *vacancies*. These will clearly upset a perfect lattice. Quite the opposite to vacancies, we can also have *extra* atoms sitting off lattice sites, between the lattice atoms. These are called *interstitials* and they will also upset the lattice periodicity. Finally, if our piece of silicon is sitting at a finite temperature, the lattice atoms will be vibrating about their mean positions giving rise to those momentum-carrying 'particles' we met in the last chapter, the *phonons*. Our current carriers will interact with these phonons, with momentum being transferred, and, again, carrier progress will be impeded. So, as we already know by now, a real piece of semiconductor does not behave like a superconductor, it has some resistance. We have just had a quick look, at the microscopic level, at some of the mechanisms that can cause this resistance. We want a microscopic understanding of mobility, and we can see that the mobility will have at least two components to consider: a lattice interaction component and an impurity interaction component. Let's get down to the microscopic level with our charge carrier and see what's going on.

We're an electron and we're wandering through a piece of crystalline silicon. Why are we wandering at all? The silicon is in thermodynamic equilibrium and there's no electric field! We are moving because we have thermal energy associated with being at a temperature T. How much energy do we have? We have $kT/2$ of energy per *degree of freedom*. That's just a fancy way of saying that for each dimension we move in (time not included) there is an associated $kT/2$ of thermal energy. As an example, a particle that can move in three dimensions (x, y, z) will have $3kT/2$ worth of thermal energy associated with its motion. Since we are able (normally) to move in any of the three dimensions within a piece of silicon, then we will have $3kT/2$ worth of thermal energy as well. That's what's powering us! We're moving around with an average speed given by

$$v_{th} = \sqrt{\frac{3kT}{m^*}} \tag{4.4}$$

where v_{th} = thermal velocity (average). This equation was obtained very simply by equating our thermal energy to our kinetic energy of $m^*v^2/2$, where m^* is an effective mass, which we need to consider since we are moving in a crystal. Throughout this chapter I will be using *average values* because in reality the velocities we are considering follow certain *distributions*, in fact the Maxwell–Boltzmann distribution that we came across in Chapter 1. I don't want to get involved with the details of these statistics because there's not enough payback, but bear in mind that the average values we use are the averages of a statistical distribution given by Maxwell–Boltzmann statistics. OK, so we're going along minding our own business when BANG! we get clobbered by a phonon, and those things carry some momentum! Momentarily we are stopped in our tracks, our velocity is zero, but we then pick ourselves up and carry on again in a *random direction* to our original path. This randomness is quite important. It means as many particles are moving left as right, up as down, in as out. There is therefore no *net* current flow under these conditions. You cannot connect a torch bulb to a piece of material under thermal equilibrium conditions and expect the bulb to light up because there is no net current flow through a material in thermal equilibrium. We start moving off again and again we get clobbered; this time it's the disturbance in the perfect lattice potential field caused by an ionized impurity. I think it might have been an oxygen atom; those things get everywhere. I hope you get the picture. The free carriers in a semiconductor appear to move around randomly (just like the smoke particles or pollen grains in Brownian motion) due to their interactions with impurities, defects, vacancies, phonons and anything else that upsets the perfect crystalline lattice structure. Enough of the microscopic musings, let's get back to some hard numbers.

Calculation

Calculate the velocity of an electron in a piece of n-type silicon due to its thermal energy at room-temperature, and due to the application of an electric field of 1000 V/m across the piece of silicon. Comment on the effect of the electric field.

Simply plug the numbers into equation (4.4) to get a thermal velocity v_{th} of $1.08 \times 10^5 \, \mathrm{m \, s^{-1}}$ using an effective electron mass of $1.18m_0$.

Use equation (4.3) with an electron mobility of $0.15 \, \mathrm{m^2 V^{-1} s^{-1}}$ (n-type silicon) to get a drift velocity v_d of $150 \, \mathrm{m s^{-1}}$.

The drift velocity is seen to be almost insignificant compared to the 'natural' thermal velocity. The 'added' velocity due to the electric field is thus a very small perturbation on the electron's (or hole's) normal thermal velocity. The electron's 'natural' velocity is hardly perturbed at all by the application of electric fields typical in semiconductor devices. The time between collisions is therefore also unperturbed by the electric field. We shall use this result in the next section.

We still want a microscopic understanding of the mobility, in other words we want an equation with mobility on one side and some fundamental quantities on the other. We can do this by going back to our electron undergoing its collisions in the silicon crystal. We can then bring in the fundamental quantities by asking, how long does the electron go in time before a collision, or, equally valid, how far does it go in space before a collision? The average *time* between collisions is called the *relaxation time* or the *mean free time*, τ. The average *distance* between collisions is known as the *mean free path*. We are now ready to derive a microscopic equation for the mobility! The drift velocity will be given by the product of carrier acceleration (due to the applied electric field) and the time for which the acceleration is acting (the mean free time). The acceleration is simply the force acting divided by the carrier mass where the force acting is given by equation (4.1). The drift velocity can therefore be written as

$$v_d = \frac{\mathbf{F}\tau}{m} \tag{4.5}$$

where

$\mathbf{F}$ = force acting due to electric field,
τ = the relaxation time.

If we now substitute for $\mathbf{F}$ using equation (4.1) we get

$$v_d = \frac{q\mathbf{E}\tau}{m} = \mu E \tag{4.6}$$

where we have also used (4.3). Equation (4.6) thus leads us to the 'microscopic' equation for the mobility:

$$\mu = \frac{q\tau}{m^*} \tag{4.7}$$

Calculation

Calculate the mean free time and mean free path for electrons in a piece of n-type silicon and for holes in a piece of p-type silicon.

For n-type silicon, as before, we will use an electron mobility of $0.15\,\text{m}^2\text{V}^{-1}\text{s}^{-1}$ and an electron effective mass of $1.18m_0$. Substitution into equation (4.7) gives a mean free time of $10^{-12}\,\text{s}$. To get the mean free path, l, we multiply the thermal velocity (found earlier for electrons) by the mean free time, giving

$$l = v_{th}\tau = 10^{-7}\,\text{metre}$$

For holes the procedure is exactly the same except we need to plug in some different values such as mobility $0.0458\,\text{m}^2\text{V}^{-1}\text{s}^{-1}$ and effective mass $0.59m_0$. You should get

mean free time $= 1.54 \times 10^{-13}\,\mathrm{s}$
thermal velocity $= 1.52 \times 10^{5}\,\mathrm{m\,s^{-1}}$
hence, mean free path $= 2.34 \times 10^{-8}\,\mathrm{m}$

4.3 Saturated drift velocity

Equation (4.3) implies that we can make a carrier go as fast as we like just by increasing the electric field. Unfortunately for those of us that work with silicon technology, this is not the case. Figure 4.2 shows how the electron drift velocity *saturates* at high electric fields. If you look at the drift velocity, v_d, at which the saturation seems to start you can see that it is pretty close to the carrier thermal velocity, v_{th}, that we calculated earlier. As we try to increase carrier velocity by increasing the electric field beyond this point, we find that the expected energy increase goes into heating up the lattice, rather than into the electron's own kinetic energy. So you can see that we have a fundamental problem here when it comes to trying to make *fast* devices. Real devices often have both high-field and low-field regions in their construction. In the low-field regions the carrier mobility will be a major limitation to device speed, whereas in the high-field regions, the limitation is the saturated drift velocity. In *real* devices speed is also hampered by *parasitics*, unwanted stray resistances, inductances and capacitances that are an unavoidable consequence of the device *structure* and packaging.

4.4 Mobility variation with temperature

It was mentioned in Section 4.2 that the mobility will be made up of *at least* two components associated with lattice (scattering) interactions and impurity (scattering) interactions. If the mobility is made up of several components say μ_1, μ_2, μ_3, then the total mobility μ is given by

$$\mu = \left(\frac{1}{\mu_1} + \frac{1}{\mu_2} + \frac{1}{\mu_3} \right)^{-1} \tag{4.8}$$

If we consider just two components due to lattice scattering, μ_L, and ionized impurity scattering, μ_I, then the mobility is given by

$$\mu = \left(\frac{1}{\mu_I} + \frac{1}{\mu_L} \right)^{-1} \tag{4.9}$$

As the lattice warms up, the lattice scattering component will increase, and the mobility will *decrease*. Experimentally, it is found that the mobility decrease often follows a $T^{-1.5}$ power law:

$$\mu_L = C_1 \times T^{-\frac{3}{2}} \tag{4.10}$$

where C_1 is a constant.

As we cool a piece of semiconductor down, the lattice scattering becomes less important, but now the carriers are more likely to be scattered by ionized

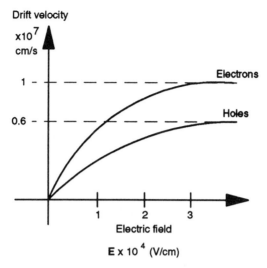

Figure 4.2 Saturation drift velocity.

impurities, since the carriers are moving more slowly. It is found that the ionized impurity scattering often follows a $T^{1.5}$ power law giving:

$$\mu_I = C_2 \times T^{\frac{3}{2}} \tag{4.11}$$

Figure 4.3 shows the variation in carrier mobility as a function of temperature due to the presence of both ionized impurity and lattice scattering effects. Note that there is a peak in the mobility curve at a temperature that depends on the number density of ionized impurities. In practice, the number density of ionized impurities is simply the (shallow) doping-level density. Highly doped samples will therefore cause more scattering, and have a lower carrier mobility, than low-doped samples. This fact is used in high-speed devices called High Electron Mobility Transistors (HEMTs) where electrons are made to move in *undoped* material, with the resulting high carrier mobilities!

4.5 A derivation of Ohm's law

Starting with our current defining equation (4.2) in one dimension:

$$I_d = nAqv_d$$

first convert to current density $(A\,m^{-2})$ to obtain

$$J_x = nqv_d \tag{4.12}$$

where J_x is the current density in the x-direction. Rewrite equation (4.12) using equations (4.3) and (4.7) to get

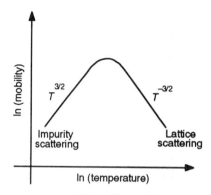

Figure 4.3   Mobility as a function of temperature.

$$J_x = \left(\frac{n\tau q^2}{m^*}\right)E_x \tag{4.13}$$

where we now have a relationship between *current density* and *electric field*. This equation is in fact just Ohm's law in a slightly different form. In order to see that equation (4.13) really is just Ohm's law we need to introduce the *conductivity*. The (electrical) conductivity of a material is simply the proportionality factor between the current density and the applied electric field, which in one dimension can be written as

$$J_x = \sigma_x E_x \tag{4.14}$$

where σ_x is the electrical conductivity with units of Sm^{-1}. The reciprocal of conductivity is called the *resistivity*:

$$\rho_x = \frac{1}{\sigma_x} \tag{4.15}$$

where ρ_x is the electrical resistivity with units of Ωm.

You can see immediately from equation (4.13) that the conductivity, in terms of fundamental quantities, is given by

$$\sigma = \left(\frac{nq^2\tau}{m^*}\right) \tag{4.16}$$

Finally, convince yourself by referring to Figure 4.4 that the *resistance* of a block of material of *resistivity* ρ across the two end faces of area A is just

$$R = \frac{\rho l}{A} \tag{4.17}$$

where you can resort to analysis of the units for confirmation.

We can now transform equation (4.13) into the more familiar Ohm's law. Starting with,

$$J_x = \left(\frac{n^2 q\tau}{m^*}\right)E_x = \sigma E_x$$

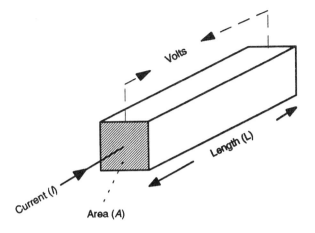

Figure 4.4 Diagram for resistivity.

write down the current density as current divided by area, and electric field as potential divided by length to obtain

$$\frac{I_x}{A} = \sigma \frac{V_x}{L_x}$$

Rearrange to get

$$V_x = \frac{L_x I_x}{A} \rho$$

Finally, use equation (4.17) to obtain

$$V_x = I_x R \tag{4.18}$$

which is the form of Ohm's law with which you'll be most familiar.

4.6 Drift current equations

We have now covered enough material to look at the drift current equations with some numerical examples.

Considering intrinsic semiconductors first, we can use equations (4.7) and (4.13) to write down the electron current density as

$$J_n = nq E \mu_n \tag{4.19}$$

where μ_n = electron mobility.

For holes the current density will be

$$J_p = pq E \mu_p \tag{4.20}$$

where μ_p = hole mobility.

Since we are dealing with intrinsic material $n_i = n = p$ and so the total current density will be given by

$$J = J_n + J_p = En_iq(\mu_n + \mu_p) \qquad (4.21)$$

giving an intrinsic conductivity of

$$\sigma_i = \frac{J}{E} = n_iq(\mu_n + \mu_p) \qquad (4.22)$$

Calculation

Calculate the intrinsic conductivity of silicon at room-temperature.

Using equation (4.22) with $n_i = 1.4 \times 10^{10}\,cm^{-3}$, $\mu_n = 1500\,cm^2\,V^{-1}s^{-1}$, and $\mu_p = 458\,cm^2\,V^{-1}s^{-1}$, we obtain

$$\sigma_i = 1.4 \times 10^{16} \times 1.6 \times 10^{-19}\ (0.15 + 0.046) = 4.4 \times 10^{-4}\,Sm^{-1}$$

Taking the reciprocal gives us the intrinsic resistivity of silicon at room-temperature:

$$\rho_i = \frac{1}{\sigma_i} = 2.27 \times 10^3\,\Omega m$$

Let's go on to calculate the *resistance* of a piece of intrinsic silicon. If we consider a cube of silicon of length 1 cm on a side, then we can use equation (4.17) to calculate the resistance across two end faces given the resitivity we have just found above. Substituting values into equation (4.17) gives a resistance of $2.27 \times 10^5\,\Omega$!! Our intrinsic block of silicon is *very* resistive.

For *extrinsic* semiconductors the current density is again due to electron and hole flow:

$$J = nqE\mu_n + pqE\mu_p \qquad (4.23)$$

However, for n-type semiconductor material where $n \gg p$ we obtain

$$\sigma_n = qN_d\mu_n \qquad (4.24)$$

where N_d is the shallow donor concentration.

Similarly, for a p-type semiconductor where $p \gg n$ we get

$$\sigma_p = qN_a\mu_p \qquad (4.25)$$

with N_a being the shallow acceptor concentration.

Calculation

A sample of intrinsic silicon of volume $1\,cm^3$ is illuminated by a light pulse of power 1 mW for 10 µs from a helium–neon laser (wavelength 632.8 nm). Assume the photons are absorbed uniformly throughout the silicon and that each photon creates one electron–hole pair. By how much does the conductivity change after irradiation? Assume no carrier recombination.

The conductivity without illumination is given by equation (4.21):

$$\sigma_i = n_i q (\mu_n + \mu_p)$$

With illumination the carrier concentration will increase from n_i to $n_i + \delta n$, where δn is the number of electron–hole pairs per unit volume created by the light pulse. Hence the change in conductivity is given by

$$\frac{\sigma_{light}}{\sigma_{dark}} = \frac{n_i + \delta n}{n_i}$$

The photon energy is given by hc/λ, with h = Planck's constant and c = velocity of light *in-vacuo*. The total number of electron–hole pairs produced will be given by

$$\delta n = (\text{light pulse power} \times \text{time})/(\text{energy per photon})$$

so that

$$\delta n = \frac{(10^{-3} \times 10 \times 10^{-6} \times 0.6328)}{(1.6 \times 10^{-19} \times 1.24 \times 10^{-6})} \, \text{m}^{-3}$$

Hence,

$$\delta n = 3.2 \times 10^{16} \, \text{m}^{-3}$$

And finally the conductivity change is found to be

$$\frac{\sigma_{light}}{\sigma_{dark}} = \frac{3.2 \times 10^{16} + 1.4 \times 10^{16}}{1.4 \times 10^{16}} = 3.3$$

So the conductivity of the silicon sample is just over three times greater immediately after the light pulse. This example shows how a piece of intrinsic silicon could be used as a photoconductive detector.

Question

Why does the resistivity of a metal *increase* with increasing temperature, whereas the resistivity of a semiconductor *decreases* with increasing temperature?

This one is a little more subtle. If we use the reciprocal of equation (4.24) as our basis for resistivity:

$$\rho = \frac{1}{q n \mu}$$

then we can start to see what the difference between metals and semiconductors might be when the temperature is increased.

For a metal, the number density of conduction electrons hardly changes with temperature at all. However, we know that the metal's mobility must decrease with increasing temperature, because of the increased lattice vibrations. So if the carrier number density n is not changing, but the mobility μ is decreasing, then overall the metal resistivity will *increase* with increasing temperature. For the semiconductor we know the carrier number density will *increase* with increasing temperature (more electron–hole pairs will be generated), but the mobility will *decrease* with increasing temperature for the same reason as given above. At first glance it may not be obvious what the resistivity should do; however, if you look at some Chapter 3 material you'll see that the carrier number density increases *exponentially* with increasing temperature. From Section 4.4 of this chapter we know that the mobility decreases with temperature following a $T^{-1.5}$ power law. The exponential 'beats' the straight power to 1.5 and so the outcome is that the semiconductor resistivity will *decrease* with increasing temperature. Again, this fact is used in a real semiconductor device called a *thermistor*, which is used as a temperature sensing element.

4.7 Semiconductor band diagrams with an electric field present

We've already seen simple energy-band diagrams for p- and n-type semiconductors in Chapter 3 (Figure 3.11). These diagrams were drawn for a semiconductor in thermal equilibrium with no external electric field applied. In real semiconductor devices there will be electric fields present and we need to know how to accommodate them in our band diagrams.

Figure 4.5 shows how the energy bands are modified when an electric field is applied. Without bothering about any maths for the moment, just look at the pictures and see if they make any sense. You'll see as the book progresses that I put a lot of weight on having the right picture. You don't actually need to have lots of maths, or equations, at your fingertips provided you have the right physical picture, and can draw the correct band diagram. For this reason I will look at real devices from a band-diagram approach, rather than a mathematical approach. After all a picture is worth . . . .

Let's get back to Figure 4.5. Look at the n-type semiconductor with the applied field. There will be a force on the electrons (majority carriers) and it will be to the right, as seen in the diagram. Look at how the bands have moved: they slope down to the right and so an electron wanting to be at the lowest energy in the conduction band will move to the right. How about the hole picture? The lowest hole energy is at the top of the valence band and so holes will tend to move up the valence band slope, from right to left, as shown in the diagram. Obviously, currents are flowing, we've connected a battery across a piece of n-type semiconductor. Because the currents are flowing due to an electric field we are talking about drift currents. Now we come to a useful mnemonic: under *drift* conditions holes float and electrons sink. Look at the diagram again, the electrons fall down the conduction band slope and will tend to move to the slope's lowest point (like a ball rolling down a hill). The holes 'float' to the highest point of the

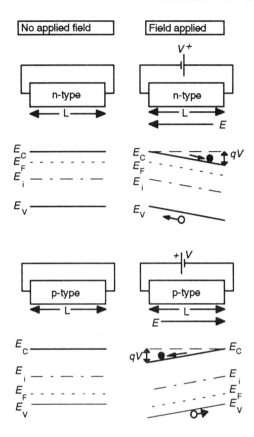

Figure 4.5   Band movement in an electric field.

valence band slope. But beware! This is for *drift conditions only*!!! Note the directions of the electric fields in the diagrams and go through the same arguments as above for the p-type case. The physical pictures are clearly reasonable, but that doesn't make them correct. We have to go to the maths to see if the pictures hold up to more rigorous analysis.

Because signs are going to become important now, I will be a little more careful in how I write down the equations. Vectors will have a little bar over the appropriate letter and I will put $+q$ or $-q$ when dealing with holes and electrons. Don't worry too much about things being vectors at this stage, especially since we work mostly in one dimension anyway. However, as you get more confident with vectors (by the time you've covered the quantum mechanics chapters) you can come back to the equations and see what they mean with more understanding.

You can see from Figure 4.5 that I have drawn the conduction and valence bands with a certain slope, and I have said that the 'depth' of this slope from one end to the other is qV, where V is just the battery voltage. We need to see that this is correct. We'll start off with an equation that you'll all be happy with:

$$WORK = \textbf{FORCE} \times DISTANCE \qquad (4.26)$$

I know you'll be happy with that because I was at high-school level as well. It was only when I came to write this book that I became a little unhappy and realized the power of that equation for the first time!

What force are we talking about? It's just the electrostatic force, the very first equation of this chapter. And the distance? That's just the length of the piece of semiconductor, L. So, mathematically,

$$WORK = -q\bar{E} \times L \qquad (4.27)$$

The total work done will be the total gain in energy in moving an electron from the right-hand side of the piece of semiconductor to the left-hand side. Remember the y-axis on these diagrams is electron energy so whatever number we come up with, that will be the 'depth' of the slope on the conduction band (refer to Figure 4.5 again).

The electric field is simply the potential across the sample, V, divided by the sample length, L. Equation (4.27) can therefore be rewritten as

$$GAIN\ IN\ ENERGY = -qV \qquad (4.28)$$

And this is the depth of the slope, as we expected. What is the slope of a band? Picking either the conduction or valence band, or in fact the slope of the intrinsic Fermi-level or Fermi-level itself, we see that the slope is

$$SLOPE = -\frac{qV}{L} = -q\bar{E} = \textbf{FORCE ON ELECTRON} \qquad (4.29)$$

So the *steepness of the slope* gives you an idea of the *force* acting on the current carriers. We shall get one last result from Figure 4.5 that we shall use when we look at the p–n junction, but before deriving this, look at the Fermi-level again. The Fermi-level for the first time in one of our diagrams is *not horizontal*. This means that the system is *not* in thermal equilibrium and therefore *currents are able to flow*.

For the last part of this section I want to find a relationship between a quantity called the electrostatic potential and the potential energy. I need this particular result for the p–n junction discussed in Chapter 6.

The force acting can be defined as minus the gradient of (differential with respect to distance) the potential energy,

$$\textbf{FORCE} = -GRADIENT\ OF\ POTENTIAL\ ENERGY = -\frac{dV}{dx} \qquad (4.30)$$

As we are considering the slope (gradient) of the potential energy we could consider the slope of the conduction band, the valence band, the Fermi-level or the intrinsic Fermi-level. For reasons that become clearer in Chapter 6 we shall use the *intrinsic Fermi-level*. The force acting is again the electrostatic force, so, in one dimension,

$$-q\textbf{E}_x = -\frac{dE_i}{dx} \qquad (4.31)$$

Hence an equation relating the electric field to the slope of the potential energy is given by

$$E_x = \frac{1}{q}\frac{dE_i}{dx} \tag{4.32}$$

We can define the *electron electrostatic potential* V_n whose negative gradient is equal to the acting electric field:

$$E_x = -\frac{dV_n}{dx} \tag{4.33}$$

Comparison of equations (4.32) and (4.33) gives a relation between the electron's electrostatic potential and the potential energy of an electron:

$$V_n = -\frac{E_i}{q} \tag{4.34}$$

We've covered quite a lot of material on carrier drift and carrier mobility, and there's an awful lot more to learn for a reasonable understanding of this topic, but for now that's enough. With the worked examples, you know enough about carrier drift to understand what's happening in the drift region of devices such as p–n junctions and transistors. What we don't know yet is what happens in the low-field regions where diffusion of carriers predominates. That is the subject matter of the second part of this chapter.

4.8 Carrier diffusion

If we have carrier movement in a specific direction, then we have a rate of flow of charge in that direction, i.e. a current. It doesn't matter what mechanism is causing the carrier movement, the outcome is still the same, a current flows. The mechanism we're looking at this time is called diffusion. This is just the same as the diffusion of gases or the diffusion of dopants into semiconductors at elevated temperatures. This being the case, the *form of the equations* will be the same for all these situations, and we shall 'borrow' some gas law equations to describe carrier diffusion.

You know that if you have a concentration gradient in a gas, say a lot of chlorine in a corner of a chemistry laboratory, then there is a tendency for the gas to diffuse away from the area of high concentration. The gas does this to equalize out the concentration of gas in the room. But why does it do this? If we look at the basic *ideal gas equation*:

$$P = nkT \tag{4.35}$$

where

P = pressure of the gas,
n = number of gas molecules per unit volume,

then we can differentiate (4.35) to see the effect of a concentration gradient. In one dimension, x, we get

$$\frac{dn}{dx} = \frac{1}{kT}\frac{dP}{dx}$$

(4.36)

so the concentration gradient produces a pressure gradient and it is this pressure gradient that produces the force causing the gas to expand. Things will obviously come to a stop when we reach an equilibrium condition with a uniform concentration of gas throughout. What has all this talk about ideal gases got to do with charge carriers in semiconductors? We can produce carrier concentration gradients in semiconductors by several mechanisms, and just as with the gas case, the carriers will diffuse away from high concentration regions to low-concentration regions. As the carriers diffuse, a current flows, which is called the *diffusion current*. How can we produce a carrier concentration gradient in a semiconductor? We can *inject* carriers into semiconductors using metallic or semiconductor contacts. This will produce an increased concentration of carriers near the contact and the carriers will diffuse away from this region. Another way of producing a concentration gradient is to use a beam of light, see Figure 4.6. In this case the incident photons have an energy greater than the semiconductor's band gap and so they produce electron–hole pairs where the photons are absorbed. Locally there will be an increased density of electron–hole pairs which will diffuse away from the high-concentration region in which they are created.

If we are to calculate diffusion currents we need some diffusion equations. How do we formulate these? Figure 4.7 shows an arbitrary concentration of electrons in the x-direction. Since the electrons will move from high to low concentration they will move towards positive x, giving a conventional current flow in the negative x-direction. The rate at which carriers diffuse down the concentration gradient is found to be proportional to the concentration gradient, i.e.,

$$FLUX = -D\frac{dn}{dx}$$

(4.37)

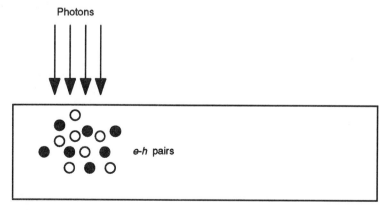

Figure 4.6 Carrier diffusion.

where

    flux = number of carriers crossing unit area per unit time, so the units of flux are $(m^{-2}s^{-1})$,
    D = the *diffusion coefficient*, units are (m^2s^{-1}).

We need to be able to derive equation (4.37) in order to understand the underlying physics.

4.9 The flux equation

Figure 4.8 shows the situation we'll be considering with some completely arbitrary electron concentration gradient. We are fortunate in that we've already covered a couple of key concepts that we need. The mean free path, and the mean free time, were discussed in the drift section of this chapter.

Since we are at a finite temperature, the electrons will have a random thermal motion with a thermal velocity v_{th} and a mean free path l. Any electrons at $x = -l$ have equal chances (remember the randomness) of moving left or right, and in a mean free time τ, half of them will cross the plane at $x = 0$. So the average rate of electron flow per unit area (*the flux*), F_1, of electrons crossing the $x = 0$ plane from the left is

$$F_1 = \frac{n(-l)l}{2\tau} = \frac{n(-l)v_{th}}{2} \qquad (4.38)$$

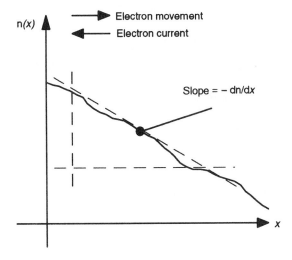

Figure 4.7 Electron diffusion.

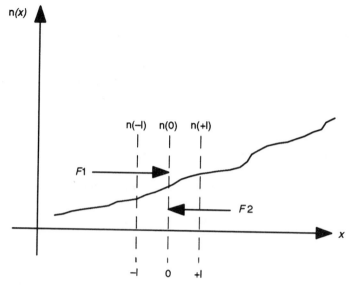

Figure 4.8 Flux diagram.

where

$$n(-l) = \text{the carrier concentration at } x = -l$$

Using exactly the same reasoning the flux of electrons crossing the $x = 0$ plane from the right is

$$F_2 = \frac{n(l)v_{th}}{2} \tag{4.39}$$

So the *net rate of carrier flow* from left to right is

$$F_{NET} = F_1 - F_2 = v_{th}\frac{[n(-l) - n(l)]}{2} \tag{4.40}$$

Some differential calculus allows us to approximate $n(-l)$ and $n(l)$ according to

$$n(-l) \approx n(0) - l\frac{dn}{dx} \tag{4.41}$$

$$n(+l) \approx n(0) + l\frac{dn}{dx} \tag{4.42}$$

If we substitute (4.41) and (4.42) into (4.40) we obtain

$$F_{NET} = -v_{th}l\frac{dn}{dx} = -D_n\frac{dn}{dx} \tag{4.43}$$

where $D_n = v_{th}l$ is the electron diffusion coefficient $(\text{m}^2\text{s}^{-1})$.

While we're at it we may as well write down the current density that equation (4.43) gives us. Since we're talking about fluxes, the flow per unit area per second, we only need to multiply by the charge per carrier to get the current density:

$$J_n = -q \times -D_n \frac{dn}{dx} = qD_n \frac{dn}{dx} \text{ Am}^{-2} \tag{4.44}$$

The same argument applied to a hole concentration gradient gives

$$J_p = +q \times -D_p \frac{dp}{dx} = -qD_p \frac{dp}{dx} \text{ Am}^{-2} \tag{4.45}$$

4.10 The Einstein relation

The Einstein relation is a fundamental equation of semiconductor physics which links the diffusion coefficient, D, to the carrier mobility μ. We shall derive the Einstein relation for electrons and holes separately so you have a full understanding of the derivation. Starting with electrons first, consider Figure 4.9. We have an electron concentration gradient increasing with increasing x. We know that a concentration gradient gives rise to a 'pressure' gradient and that the force acting on the element of width δx will be towards the negative x-direction and will be given by

$$\textbf{Force on element} = -\frac{dP}{dx}\delta x \text{ per unit area} \tag{4.46}$$

The number of charge carriers per unit area in the element is just $n\delta x$. Hence, the force *per carrier* in the element δx is given by

$$\textbf{Force per carrier} = -\frac{1}{n}\frac{dP}{dx} \tag{4.47}$$

From equation (4.36) we see that

$$\frac{dP}{dx} = kT\frac{dn}{dx} \tag{4.48}$$

which, substituting into equation (4.47), gives

$$\textbf{Force per carrier in element} = -\frac{1}{n}kT\frac{dn}{dx} \tag{4.49}$$

The thermal velocity is related to the force acting by

$$v_{th} = \frac{\tau}{m^*} \times \textbf{Force} \tag{4.50}$$

If you're unhappy with (4.50), just go through the steps we used in the section on carrier drift:

$$v_{th} = \mu E \tag{4.3}$$
$$\textbf{Force} = qE \tag{4.1}$$
$$\mu = q\frac{\tau}{m^*} \tag{4.7}$$

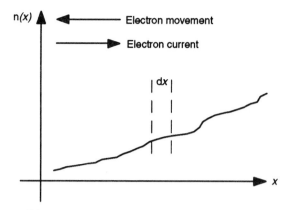

Figure 4.9 Electron diffusion.

Hence,

$$v_{th} = \frac{\mu \mathbf{Force}}{q} = \frac{\tau}{m^*} \times \mathbf{Force}$$

The diffusion current density due to the electrons is

$$J_n = n(-q)v_{th} \tag{4.51}$$

Substitute for v_{th} and for the force per carrier from equation (4.49) to get

$$J_n = \left(\frac{q\tau}{m^*}\right)kT\frac{dn}{dx} \tag{4.52}$$

Substituting the mobility for its fundamental quantities gives

$$J_n = \mu kT\frac{dn}{dx} \tag{4.53}$$

But from equation (4.44),

$$J_{DIFF} = qD_n\frac{dn}{dx}$$

So equating (4.44) and (4.53) we finally get the Einstein relation

$$D_n = \frac{\mu_n kT}{q} \quad \text{EINSTEIN RELATION} \tag{4.54}$$

Although long-winded, the derivation is in fact pretty straightforward; let's go through the same thing for holes this time, missing out a few of the obvious steps. The relevant diagram is Figure 4.10.

The force per carrier in element δx is given by

$$\mathbf{Force} = -\frac{1}{p}kT\frac{dp}{dx} \quad \{\text{compare with (4.49)}\} \tag{4.55}$$

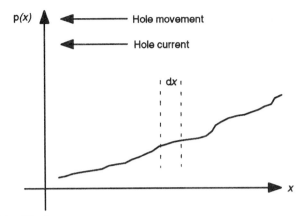

Figure 4.10 Hole diffusion.

The hole current density is given by

$$J_p = p(+q)v_{th} \quad \{\text{compare with (4.51)}\}$$

Hence,
(4.56)

$$J_p = -\frac{q\tau}{m^*}kT\frac{dp}{dx} \quad \{\text{compare with (4.52)}\}$$
(4.57)

And equating (4.57) to (4.55) again gives the Einstein relation,

$$D_p = \frac{\mu_p kT}{q} \quad \text{EINSTEIN RELATION}$$
(4.58)

Calculation

For silicon, calculate values for the electron and hole diffusion coefficients given their respective carrier mobilities.

Simply substitute into equations (4.54) and (4.58) using the usual mobilities of $0.15\,m^2/V\,s$ for electrons and $0.046\,m^2/V\,s$ for holes. Note that kT/q has a value of $25\,mV$ (prove this to yourself).

Making the substitutions you should get

$D_n = 3.75 \times 10^{-3}\,m^2/s$
$D_p = 1.15 \times 10^{-3}\,m^2/s$

So electrons drift *and* diffuse faster than holes in silicon (this is true for GaAs as well). This being the case, if its speed you're after, make devices that depend on the movement of electrons rather than holes!

4.11 Total current density

When an electric field is present in addition to a carrier concentration gradient, both *drift and diffusion* currents will flow. The total current density in the x-direction will therefore be given by the sum of both components:

$$J_n = q\mu_n n \mathbf{E}_x + qD_n \frac{dn}{dx} \tag{4.59}$$

$$J_p = q\mu_p p \mathbf{E}_x - qD_p \frac{dp}{dx} \tag{4.60}$$

So the *total* current density is just the sum of (4.59) and (4.60):

$$J_{\text{TOTAL}} = q\mathbf{E}_x(\mu_n n + \mu_p p) + q\left(D_n \frac{dn}{dx} - D_p \frac{dp}{dx}\right) \tag{4.61}$$

The above equations hold good for low-field conditions but at high electric fields where we run into velocity saturation, $\mu_n \mathbf{E}_x$ and $\mu_p \mathbf{E}_x$ must be replaced by the appropriate saturation drift velocity.

4.12 Carrier recombination and diffusion length

The calculation we did in Section 4.6 (where we illuminated a silicon sample with a helium–neon laser) told us to ignore carrier recombination. In this section we shall see what carrier recombination entails, and how this leads to a characteristic length over which carriers can diffuse before recombining – *the diffusion length*.

When we illuminated our silicon sample we produced excess carriers. Suppose we have a piece of n-type silicon doped to $10^{15}\,\text{cm}^{-3}$ and illuminate it with light that produces $10^{14}\,\text{cm}^{-3}$ electron–hole pairs. The electron concentration (the majority carriers) hardly changes, but the hole concentration (the minority carriers) goes from a very low value of $1.96 \times 10^5\,\text{cm}^{-3}$ to $10^{14}\,\text{cm}^{-3}$. By this mechanism a lot of excess holes are produced. These minority carriers find themselves surrounded by a very high concentration of majority carriers and will readily recombine with them. It is found that the recombination rate is proportional to the *excess* carrier density, so if δp is the excess hole density, then

$$\frac{d\delta p}{dt} = -\frac{1}{\tau_p}\delta p \tag{4.62}$$

which defines the lifetime, τ_p, of the holes. The solution to the differential equation (4.62) should be well-known to you. The *form* of the equation is the same as for *radioactive decay*. Whenever you get a quantity changing at a rate that is proportional to the amount of the quantity, you get this solution. The solution is

$$\delta p(t) = \delta p(0)\exp\left(-\frac{t}{\tau}\right) \tag{4.63}$$

which gives the excess hole concentration as a function of time, where,

$$\delta p(0) = \text{excess hole concentration when } t = 0$$

While the light source is on excess carriers are being generated; they are also recombining all the time the light is on as well. Clearly, when the light source is removed the generation of excess carriers ceases, and the excess carrier concentration will decay exponentially with time. The form of the equations is just the same for the electrons:

$$\frac{d\delta n}{dt} = -\frac{1}{\tau_n}\delta n \qquad (4.64)$$

with

$$\delta n(t) = \delta n(0)\exp\left(-\frac{t}{\tau}\right)$$

but the exponential decrease in the electron concentration is a lot less dramatic since the number density decays from only $1.1 \times 10^{15}\,\text{cm}^{-3}$ to $10^{15}\,\text{cm}^{-3}$, whereas the hole concentration decays from $10^{14}\,\text{cm}^{-3}$ to $1.96 \times 10^{5}\,\text{cm}^{-3}$ (the equilibrium minority carrier concentration).

In silicon (and other indirect band-gap semiconductors) the recombination of electrons with holes occurs via impurity levels in the forbidden energy gap. When these impurity levels are being considered as recombination centres they are often called traps, or recombination centres, or lifetime killers (remember I said there were a lot of different names for impurities in silicon!). It turns out that *the most effective* recombination centres lie near the middle of the semiconductor's band gap. Such centres remove carriers by first trapping a hole and then an electron (or vice-versa depending on the material doping and the energy position of the trap), thus removing an electron–hole pair in the process. The analogy that is often given is that these recombination centres act like 'stepping stones' for carriers between the bands. I must repeat myself and spell it out again since it is so important: *for silicon this process is how electron–hole recombination takes place.* The direct recombination of an electron with a hole across the band gap is an extremely rare event in silicon, and in other *indirect gap* semiconductors. For silicon, the impurities gold (Au) and platinum (Pt), introduce centres near the middle of the gap and these elements therefore act as very effective *lifetime killers*. Lifetime killers, as the name suggests, seriously reduce the minority carrier lifetime, i.e. they are very effective in 'mopping up' any excess minority carriers. Since a large minority carrier lifetime in the base region of a bipolar transistor is one factor that reduces transistor speed, it is clear that a lifetime killer in the transistor base will speed the transistor up. Unfortunately, it also reduces the *current gain* of the transistor. This example suggests that being able to control carrier lifetime (by introducing lifetime killers) is of great importance when designing semiconductor devices.

If we now return to the production of excess minority carriers in a semiconductor let's think about *how far* these carriers will *diffuse* before recombining. The characteristic length over which minority carriers will diffuse before recombining with majority carriers is called the *diffusion length*. The quantity $(D\tau)^{0.5}$, which contains the diffusion coefficient and the minority carrier lifetime, has the units of length, and it is in fact the *diffusion length* of minority carriers in a semiconductor. The excess carriers are found to decay exponentially with diffusion *distance*, as well as with time, so we can write down equations of a similar form:

$$\delta n(x) = \delta n(0) \exp\left(-\frac{x}{L_n}\right)$$

and,

$$\delta p(x) = \delta p(0) \exp\left(-\frac{x}{L_p}\right)$$

where

$$L_p = \sqrt{D_p \tau_p} = \text{hole diffusion length}$$
$$L_n = \sqrt{D_n \tau_n} = \text{electron diffusion length}$$

Calculation

Calculate the electron and hole diffusion lengths in silicon given a minority carrier lifetime of 10^{-8} s. In Section 4.10 we calculated

$D_n = 3.75 \times 10^{-3}$ m^2/Vs
$D_p = 1.15 \times 10^{-3}$ m^2/Vs

Hence using equations (4.66) and (4.67) you should obtain

$L_n = 6.12 \times 10^{-6}$ m
$L_p = 3.4 \times 10^{-6}$ m

Think *physically* about what the work in this section tells us. It says that in a diffusion length the excess minority carrier concentration falls by 37% (1/e). So for electrons in the above calculation, the excess concentration falls by 37% over a distance of six microns, and for holes over a distance of 3.4 microns. It is a property of exponential decay that we could pick any two points, separated by a diffusion length along a piece of semiconductor in which carriers are diffusing, and see this fall-off of 37% in the excess carrier concentration.

Let's carry out a straightforward calculation involving the diffusion length.

Calculation

A hole diffusion current of 7 mA is injected into an n-type silicon device of cross-section 1 mm^2 and the excess hole concentration falls off exponentially with distance from the injecting contact. If the diffusion coefficient is $1.15 \times 10^{-3}\,\text{m}^2/\text{s}$ and 0.3 mm from the contact the excess hole concentration is 23% of that at the contact, calculate the excess hole concentration at the contact.

I really like this question, not so much because it teaches us a lot of semiconductor physics, but for the reason that it shows us an interesting fact concerning the exponential function.

The information given is clearly pointing us towards calculating the flux and the diffusion length, so let's start with these:

$$FLUX = \frac{7 \times 10^{-3}}{1.6 \times 10^{-19} \times 10^{-6}} = 4.375 \times 10^{22}\,\text{m}^{-2}\text{s}^{-1}$$

and for the diffusion length:

$$\delta n(x) = \delta n(0)\exp\left(-\frac{x}{L_n}\right)$$

We know that for $x = 0.3$ mm, $\delta n(0) = 1$ and $\delta n(7\,\text{mm}) = 0.23$; hence,

$$0.23 = \exp\left(-\frac{0.3 \times 10^{-3}}{L_n}\right)$$

giving

$$L_n = 2.04 \times 10^{-4}\,\text{m}$$

So we have the flux, and the diffusion length; now comes the trick. From

$$\delta n(x) = \delta n(0)\exp\left(-\frac{x}{L_n}\right)$$

differentiate with respect to x to obtain

$$\frac{d\delta n(x)}{dx} = -\frac{1}{L_n}\delta n(0)\exp\left(-\frac{x}{L_n}\right)$$

and evaluate this at $x = 0$:

$$\left.\frac{d\delta n(x)}{dx}\right|_{x=0} = -\frac{\delta n(0)}{L_n}$$

This is pretty useful, since evaluating the flux at $x = 0$ gives

$$\textbf{FLUX} = -D_n\left.\frac{d\delta n}{dx}\right|_{x=0} = \frac{D_n\delta n(0)}{L_n}$$

and we have numerical results for everything but $\delta n(0)$! So, finally,

$$\delta n(0) = \frac{\textbf{FLUX} \times L_n}{D_n} = 7.76 \times 10^{21}\,\text{m}^{-3}$$

That was rather a marathon chapter for a book of this length. You should, however, now have a rather good understanding of the drift and diffusion current contributions to the total current that you will be looking at in real semiconductor devices. So although *this* chapter was rather long it will make the device chapters slightly more succinct.

CHAPTER FIVE

Gunn diode

After a long wait we finally arrive at our first device structure. If you look at the standard layout of most device texts you will see that the Gunn diode is usually considered much later on, so why am I looking at it at this stage? There are several very good reasons. Firstly, you now have enough physics background to understand the underlying mechanisms of this device. Secondly, I consider this one of the simpler devices to understand; the p–n junction is *not* a simple device to understand, although it's often the first to be dealt with in standard texts. The only other *simple* device I can think of off-hand is the photoconductor, which we shall look at in Chapter 7. What I would really like you to get out of this chapter are two things. One, the feeling that had you been around at the time, then maybe you could have figured out the device physics for yourself. Two, that the seemingly remote pure physics we have been looking at up until now really does have applications in the semiconductor industry.

I want to set the scene by giving my own feelings on this fascinating mechanism, the Gunn effect. Gunn discovered that if a sufficiently high voltage (electric field) was maintained across an ordinary piece of n-type gallium arsenide, then pulses of current, at microwave frequencies, could be produced. This I find amazing. There is no device structure as such, no p–n junctions, nothing that looks like a transistor, and yet *pulses* of current are produced from a *steady* voltage supply. I feel that this, as far as intuition goes, is as outrageous as the fact that bashing two bits of metal together (well uranium 235 anyway) can develop unimaginable power. You will see that your pure physics knowledge allows you to understand both situations, and that quite often very simple systems can show very complicated behaviour.

Before we get underway studying the device there are two points I wish to cover. The first point is although this device is called a Gunn *diode*, there is no diodic behaviour in the sense of current rectification. The term 'diode', as far as this device goes, is used in its purest sense, meaning *a two-terminal device*. The other point is that the Gunn diode is able to be used in many different *modes* of operation. As this is an introductory physics book and *not* a circuits book I will not detail these different modes of behaviour although they will be mentioned.

As is usual with a new topic, let's start off with a diagram. Figure 5.1 shows a simplified energy–momentum diagram for a piece of n-type gallium arsenide (GaAs). There's nothing in this diagram that should upset you; you've seen the basic form before when we looked at AlGaAs going from direct to indirect gap with increasing aluminium concentration. I haven't drawn the position of the donor level (which gives rise to the electrons in the conduction band minimum) because at this stage its form would cause confusion and it needs some quantum mechanics (the Heisenberg uncertainty principle) to explain what's going on. The diagram is also simplified by being symmetric about the y-axis. If you look this subject area up in other texts you may find an energy–momentum diagram that looks very *asymmetric* about the y-axis. The reason for this is that I'm looking along just one crystal plane. In the case of asymmetric diagrams look carefully at the x-axis designation; you'll find that the diagram shows *different* crystal axis directions on either side of the y-axis. What I've done, therefore, is simply to look along one crystal axis. This doesn't upset the overall discussion at all and I hope it minimizes confusion. The reason we haven't looked in detail at crystal planes and crystal axes is the same reason I leave out any topic area that may be covered in more standard texts: the payback isn't enough for the outlay in time; there's not enough simple application!

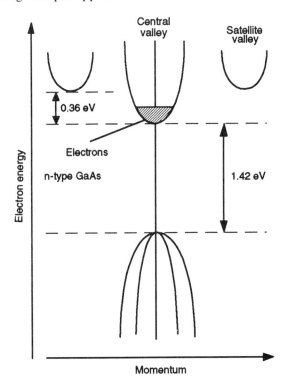

Figure 5.1   n-type gallium arsenide.

Since we're dealing with n-type GaAs we'll be most concerned with electron conduction in the conduction band, so from now on I'll be ignoring the valence band and the very few holes that are in it. Let's now just concentrate on the structure of the conduction band as it is this that gives rise to the remarkable Gunn effect. It's not surprising that the electrons want to collect in the conduction band minimum (at $k = 0$) since this is the lowest allowed energy level region in the band. However, if the two adjacent minima away from $k = 0$, called the *satellite valleys*, were only a little way in energy above the minimum at $k = 0$, then we could get some electrons in these as well. By a little way I mean that thermal energy must be sufficient to get electrons into the satellite valleys so that the difference in energy between the $k = 0$ minimum and the satellite valley minima would have to be of order kT. In Figure 5.1 you can see that at room-temperature the energy separation between the minima is much greater than kT (25 meV) and so the satellite valleys are effectively *empty of electrons*. Another key point is that although the energy separation is much greater than kT, it is still very much *less* than the GaAs band-gap energy (1.42 eV). Another couple of facts allow us to start tying all this information together to find some very strange conduction behaviour in n-type GaAs! If you compare the central valley at $k = 0$ with the satellite valleys, they have very different values of electron effective mass, namely $0.07 m_0$ for the central valley, and $0.55 m_0$ for the satellite valley. In fact if you look at the literature, you will find several different values quoted for the effective electron mass in the satellite valleys, but I'll stick with $0.55 m_0$. Since the effective masses are very different you will know that from equation (4.7) there will be very different values for the mobility. Also, the very different effective-mass values lead to very different densities of *allowed energy states* in the bands. We came across the density of states in Section 3.3 and it is discussed more fully in Appendix 1. However the key point is that the density of allowed states *increases* with *increasing* effective mass. There are therefore a lot more *allowed* energy states available in the satellite valleys than in the central ($k = 0$) valley. This greater density of allowed states in the satellite valleys is very useful; it pushes up the probability of being able to get a central valley electron *into* a satellite valley. You can see I haven't given a single equation yet. Let's try and get a physical picture of what's going on before the maths rears its ugly head.

Let's apply a big electric field to our piece of bulk n-type GaAs. By big I mean big enough to start transferring electrons from the central valley to the satellite valleys. Remember that in Section 2.2 I said electron promotion across a band gap did not usually take place by the application of small electric fields? This still holds true. For one thing we're not talking about *small* electric fields, and secondly the energy-level difference we're talking about (0.36 eV) is much smaller than the GaAs band gap. The electrons therefore transfer from a high-mobility central valley, to a low-mobility satellite valley. From equation (4.3) we see that the drift velocity must therefore change, from a *high* value at *low* fields, to a *low* value at *high* fields, contrary to intuition! From equation (4.2) we know that the above situation also means that the current falls off with increasing electric field.

This is very strange since we are producing the electric field by applying some voltage across our GaAs sample and yet increasing the electric field, i.e. increasing the voltage, *decreases* the current. In any standard resistive element increasing the voltage increases the current. What we have with the piece of bulk n-type GaAs is exactly the opposite effect; we have *negative differential resistance*!!! Don't worry about the *differential* bit; that becomes more obvious by looking at Figure 5.2. Figure 5.2(a) shows us what we already know, that at low electric fields the drift velocity is higher than at high electric fields where electron transfer from the central valley to the satellite valleys has taken place. We also know that this transfer is likely to be a smooth function (although we haven't proved this) and so a *possible* drift velocity versus electric-field curve could appear as in Figure 5.2(b). Note the region of negative differential resistance. We can't come up with *a value* for the negative resistance because it's a quantity changing along the curve. We can, however, pick any point on the curve and *differentiate at the point* to find the value of resistance. Note that depending on where you choose the point, the resistance could be either positive or negative. The Gunn mechanism relies on the *transfer* of electrons from the high-mobility central valley to the low-mobility satellite valleys. These devices are therefore also known as *transferred electron devices*, or TEDs. In order to see how current pulses are formed we need to go just one step further to see how *domains* are formed in the GaAs.

5.1 Domain formation

If we have a piece of semiconductor material behaving normally, with a positive resistance, and there is a fluctuation of charge density within the material (for whatever reason), then the charge density fluctuation is quickly 'damped out' according to

$$Q(t) = Q_0 \exp\left(\frac{-qN_D\mu_d}{\varepsilon_r\varepsilon_0}\right)t \qquad (5.1)$$

where

Q_0 = charge density at time $t = 0$,
N_D = donor density in the n-type material,
ε_r = relative permittivity of the material,
ε_0 = permittivity of free space,
μ_d = the drift mobility.

At first sight, the equation may look horrendous but we shall make one observation and then simplify the form of (5.1). If the drift mobility is positive then any charge fluctuation is going to 'die out' exponentially with time. Not only will the charge decay exponentially with time but it will do so with a *time constant* given by the reciprocal of the quantity in the parentheses. We will come back to

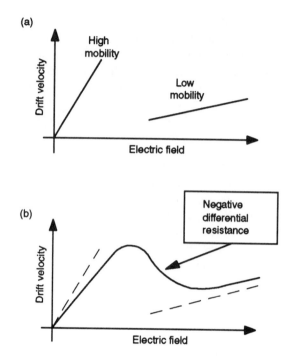

Figure 5.2   Negative differential resistance.

this in a moment. If, however, we have created a situation (by making use of the transferred electron effect) where we have a *negative* drift mobility, any charge fluctuation will not decay away with time; it will *build up* exponentially with time! This rapid build-up in charge from a small charge fluctuation leads to a region in the semiconductor containing a lot of charge, which in turn means that a large electric field exists across the same region (we shall soon see why this is so). This high-electric-field region in the semiconductor is in fact the *domain* mentioned earlier. It is this region of charge that travels from one end of the device to the other giving a *pulse* of current when it reaches the ohmic contact at the end of the device!

That describes, in a hand-waving way, the basic action of the Gunn diode. It's now time to fill in the details.

Let's return to the *time constant*:

$$\left(\frac{-qN_{\mathrm{D}}\mu_{\mathrm{d}}}{\varepsilon_{\mathrm{r}}\varepsilon_0}\right)$$

Since the exponential factor is always dimensionless we know that the time constant τ_{d} will be given by

$$\tau_d = \frac{\varepsilon_r \varepsilon_0}{q N_D \mu_d} \tag{5.2}$$

If we want to get technical, this time constant is better known as *the dielectric relaxation time*.

Calculation

How fast will a charge fluctuation *decay* away in a piece of GaAs doped with $10^{16} \, cm^{-3}$ donors?

First assume that all the donors are ionized so that there are 10^{16} electrons per cm^3 in the material. We now need values for the drift mobility and relative permittivity of GaAs. The *low-field* drift mobility (we're looking at *charge decay*) is about $8500 \, cm^2 \, V^{-1} s^{-1}$ and the relative permittivity is about 13.1 for GaAs. Substitute these values into equation (5.2) to obtain

$$\tau_d = 8.53 \times 10^{-14} \, s$$

Since any exponential decay will be effectively over in about five time constants, then the charge fluctuation will decay away in about $4.27 \times 10^{-13} \, s$. This is a *very* short time!!

We may as well look at the converse situation, the typical time for charge build-up. The maximum negative differential mobility in GaAs is about $-2400 \, cm^2/V \, s$, so using the same values as above:

$$\tau_d = 3 \times 10^{-13} \, s$$

Again, taking five time constants as being typical for the exponential event to be over we obtain $1.5 \times 10^{-12} \, s$ as the time for a charge fluctuation to build up into a *domain*.

From the above it is clear that equation (5.1) can be simplified to

$$Q(t) = Q_0 \exp -\left(\frac{t}{\tau_d}\right) \tag{5.3}$$

One last bit of analysis before we move on. Recall the equation for the conductivity of a piece of n-type semiconductor (equation (4.24)). This allows the dielectric relaxation time to be written in a form that contains the material conductivity:

$$\tau_d = \frac{\varepsilon_r \varepsilon_0}{\sigma_n} \tag{5.4}$$

I now want to return to the point about electric fields being associated with charges. This particular topic is so important that it needs its own section!

5.2 The differential form of Gauss's law

The equation describing the relationship between the electric field and charge density is extremely important in semiconductor physics (and electromagnetism) and we must tackle it. Extensive use will be made of this equation in studying the p–n junction and other devices (including, of course, the Gunn diode). However, as you know by now I don't like *just* equations, so the meaning of the equation will also be shown *graphically*. We shall use the graphical representation in looking at standard devices and in predicting the behaviour of less well-known devices (heterojunction bipolar transistors). Back to the equation, and I immediately hit a complication. The equation is so important, and so fundamental; there are at least four persons' names associated with the one basic equation! There are only *four* equations needed to describe electromagnetic phenomena. These are known as the *Maxwell equations*. The equation we want, relating electric fields to charge density is one of these equations. Before going any further, here it is:

$$\frac{\partial \mathbf{E}_x}{\partial x} + \frac{\partial \mathbf{E}_y}{\partial y} + \frac{\partial \mathbf{E}_z}{\partial z} = \frac{\rho}{\varepsilon_0} \tag{5.5}$$

Clearly, this is a partial differential equation in three dimensions. I've given it in three dimensions so that you can see the whole thing. Normally we only work in one dimension for simplicity and to save ink! In the equation ρ is the charge density, also called the *space-charge density* when dealing with charges in semiconductors. Equation (5.5) is a *differential form* of an equation worked out earlier by *Gauss* which is known as Gauss's law. Because the potential is simply related to the electric field by

$$\mathbf{E} = -\left(\frac{\partial V}{\partial x} + \frac{\partial V}{\partial y} + \frac{\partial V}{\partial z} \right) \tag{5.6}$$

we can derive an equation relating the potential to the space-charge density, namely,

$$\frac{\partial^2 V}{\partial x^2} + \frac{\partial^2 V}{\partial y^2} + \frac{\partial^2 V}{\partial z^2} = -\frac{\rho}{\varepsilon_0} \tag{5.7}$$

Equation (5.7) is known as *Poisson's equation*. For the case where no charge exists we get

$$\frac{\partial^2 V}{\partial x^2} + \frac{\partial^2 V}{\partial y^2} + \frac{\partial^2 V}{\partial z^2} = 0 \tag{5.8}$$

which is known as *Laplace's equation*.

The form of the above equations can be simplified, and they can be made much more succinct by using something called the vector differential operator ∇ which is discussed in the quantum mechanics chapters. If you want to see how vector equations can be simplified using this operator, have a quick look at the last two chapters. I won't introduce the operator here; all equations will be written out 'long-hand'.

If we are considering charges in a semiconductor, rather than in a vacuum, we have to change the permittivity from the free space permittivity to the permittivity of the semiconductor. Hence, for semiconductors,

$$\frac{\partial E_x}{\partial x} + \frac{\partial E_y}{\partial y} + \frac{\partial E_z}{\partial z} = \frac{\rho}{\varepsilon_0 \varepsilon_r} \tag{5.9}$$

Enough equations, it's time for pictures.

Figure 5.3(a) shows a rectangular distribution of charge density as may appear in a semiconductor under certain conditions. The charge contained within the region is negative and the density is shown as 10^{16} per cm^3, although the absolute value has no significance for the following discussion. Let's turn this charge density as a function of position into *electric field* as a function of position *without doing any calculations*. To get the electric field we have to integrate the charge density, since equation (5.9) can be rearranged to give

$$E_x = \frac{1}{\varepsilon_0 \varepsilon_r} \int \rho_x \, dx \tag{5.10}$$

in one dimension. We will integrate this charge density using a piece of paper! Get a piece of paper and put one edge along the left-hand edge of the charge density, keeping the shape of the charge density distribution covered up by the rest of the piece of paper. Now slowly pull the piece of paper to the right in little increments of distance, uncovering the shape of the charge distribution as you go. Mentally add up the incremental areas you are uncovering as you go. This is a lot easier than you might think! I'll run you through this first one and you can then work out the potential for yourself. When you start off you have got no uncovered area and so the electric field is zero (see Figure 5.3(b)). Now you move the piece of paper a little to the right, you uncover a little rectangle of area δA. Now you move the paper the same little distance to the right again, you uncover another little rectangle of area δA, giving now a total uncovered area of $2\delta A$. Proceeding in this way you should be able to convince yourself that the resulting integral (remember simple integration such as we're doing here is just finding the area under the curve) gives the triangular distribution of electric field, as shown in Figure 5.3(b). From the equations, the sign of the electric field will be the same as for the charge density, so the electric field will be negative as well. Now integrate the electric field for yourself to find the shape of the *potential* as a function of distance. Remember that from equation (5.6) we have to change the sign of **E** to get the correct sign for V, hence the solid curve in Figure 5.3(c) is the correct shape for the potential as a function of distance. Go over the above piece of work until you are happy with it. We shall use these results many times in the forthcoming chapters.

Probing further in our analysis of the Gunn effect, what I would like to do is derive equation (5.1), which I simply introduced, without proof, right at the beginning. We have one part of the jigsaw having covered the differential form of

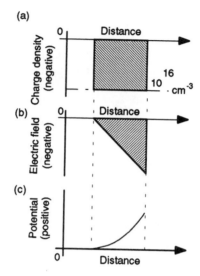

Figure 5.3 Graphical solution of Poisson's equation.

Gauss's law, but we need one other piece to finish off. What we need to derive is the charge *continuity equation*.

5.3 Charge continuity equation

I wish to derive another equation which is fundamental to many different areas of physics, the *continuity equation*. This equation applies to flow situations in which the quantity that is flowing is conserved. We shall be considering the flow of charge, charge being a quantity that is conserved. In other situations you could be considering the flow of heat, or fluids, and the *form* of the equations would be *exactly the same*. Remember the following important facts:

1 *The same physical conditions produce equations of the same form.*
2 *Equations of the same form have the same solutions.*

Look at Figure 5.4, which shows a spherical volume enclosing a positive charge Q. The *charge density* is simply given by Q/V where V is the (arbitrary) volume of the sphere. If we consider this charge to be due to a number density, p, of positive charges then we may write

$$\rho = q\mathrm{p} = \frac{Q}{V} \qquad (5.11)$$

where

ρ = the charge density (cm^{-3}),
q = the fundamental electronic charge (C).

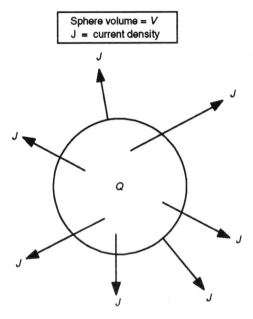

Figure 5.4 Diagram for derivation.

The loss of charge per unit time within the volume V will give rise to a current I, where

$$I = \int J \cdot dA \qquad (5.12)$$

which simply states that the total current flowing is found by integrating the *current density* over the total surface area containing the volume V (which in this case is a spherical surface). I've picked a sphere so we can use $V = 4\pi r^3/3$ and $A = 4\pi r^2$.

The loss of charge per unit time that gives rise to the current can be written as

$$-\frac{\partial}{\partial t} \int q\rho \, dV \qquad (5.13)$$

The minus sign indicates that the charge *decreases* with *increasing* time. The sign of q is positive since we are dealing with positive charges (you can see I'm being careful with the signs, otherwise we'll come unstuck a bit later). If we could make equation (5.12) an integral over volume rather than over area we could couple it easily with equation (5.13) to find a solution. Let's consider the following:

$$\int \frac{\partial}{\partial r} (J) \cdot dV \qquad (5.14)$$

For a sphere we can replace dV by $A \cdot dr$ (simply differentiate the equation for the volume of a sphere with respect to r). Substitute into equation (5.14) to get

$$\int \frac{\partial}{\partial r}(J)A \cdot dr \qquad (5.15)$$

Finally, note that the current density times the total surface area just gives the total current so that equation (5.15) becomes

$$\int \frac{\partial I}{\partial r} \cdot dr = I \qquad (5.16)$$

which fortunately means that equation (5.14) is the same as (5.12), as required (we've replaced a surface area integral by an equivalent volume integral). So if we now equate the loss of charge per unit time with the current flowing, we get

$$-\frac{\partial}{\partial t}\int q p \, dV = \int \frac{\partial}{\partial r}(J) \, dV \qquad (5.17)$$

Collecting terms gives us

$$\int \left(\frac{\partial p}{\partial t}q + \frac{\partial J}{\partial r} \right) dV = 0 \qquad (5.18)$$

But since the volume V is completely arbitrary we must have

$$q\frac{\partial p}{\partial t} + \frac{\partial J}{\partial r} = 0 \qquad (5.19)$$

or,

$$\frac{\partial p}{\partial t} = -\frac{1}{q}\frac{\partial J}{\partial r} \qquad (5.20)$$

We can simplify to one dimension and change to total differentials to obtain the simple charge continuity equation with no charge generation or charge recombination:

$$\frac{dp}{dt} = -\frac{1}{q}\frac{dJ_P}{dx} \qquad (5.21)$$

where

  p = the positive charge carrier density (cm^{-3}),
  J_P = the current density due to the flow of positive carriers $(A \, cm^{-2})$.

If we had been dealing with electrons instead of positive charges we would have used a charge of $-q$ instead of q which would give us

$$\frac{dn}{dt} = \frac{1}{q}\frac{dJ_n}{dx} \qquad (5.22)$$

At this point I suppose I'd better be honest. I've cheated! I used a specific differential property of the sphere, namely that $dV = A \cdot dr$, to get the answers I wanted. The results are in fact correct, though not rigorously derived. I did this because I didn't want to get involved with *Gauss's theorem*. For the mathematicians amongst you look up Gauss's theorem in a book on *Vector analysis* if you want to see how to do the derivation properly.

You'll be pleased to know that we've now covered enough material to work out equation (5.1) for ourselves.

5.4 The dielectric relaxation time

We shall only work in one dimension for simplicity, and we start with one of Maxwell's equations, a simplification of equation (5.10):

$$\frac{d\mathbf{E}_x}{dx} = \frac{\rho}{\varepsilon_0 \varepsilon_r} \qquad (5.23)$$

In the derivation that follows we shall only be considering negative charge carriers, electrons, because for the Gunn effect we are only considering n-type material. Multiply both sides of equation (5.23) by $dJ_x/d\mathbf{E}_x$ to obtain

$$\frac{dJ_x}{dx} = \frac{dJ_x}{d\mathbf{E}_x} \frac{\rho}{\varepsilon_0 \varepsilon_r} \qquad (5.24)$$

We again need to be very careful with signs now. Let's write down the current density flowing in the positive x-direction due to the flow of electrons in that direction:

$$J_x = -qN_D v_x \qquad (5.25)$$

where v_x is a positive quantity. Now differentiate equation (5.25) with respect to $\mathbf{E}_x$ so that we can substitute the result in equation (5.24):

$$\frac{dJ_x}{d\mathbf{E}_x} = -qN_D \frac{dv_x}{d\mathbf{E}_x} = -qN_D \mu_d \qquad (5.26)$$

where you can see that the mobility has been introduced. Substitute into equation (5.24):

$$\frac{dJ_x}{dx} = -\frac{qN_D \mu_d \rho}{\varepsilon_0 \varepsilon_r} \qquad (5.27)$$

Now introduce the charge continuity equation for negative charge flow:

$$\frac{d\rho}{dt} = \frac{dJ_x}{dx} \qquad (5.28)$$

Substitute for dJ_x/dx from equation (5.27) to obtain

$$\frac{d\rho}{dt} = -\frac{qN_D \mu_d \rho}{\varepsilon_0 \varepsilon_r} \qquad (5.29)$$

Finally, the solution of equation (5.29) gives us what we're looking for, an equation of the same form as (5.1), namely,

$$\rho = \rho_0 \exp\left(-\left(\frac{qN_D \mu_d}{\varepsilon_0 \varepsilon_r}\right)t\right) \qquad (5.30)$$

So we finally got there! Recall that for *positive* mobilities the charge density fluctuations quickly *die out*, and for *negative* differential mobilities, such as can occur in the Gunn diode at high electric fields, the charge density fluctuations *build up* very rapidly.

5.5 Operation of the TED

Having been diverted to look at some important physical equations, we can now return to the TED itself. Let's recap on what physically happens when we put enough bias on the device for domain formation to occur. If the applied bias is high enough a charge fluctuation will cause a domain (high-field region) to form very quickly. This domain travels towards the anode (positive end) with a velocity v_d until it finally reaches the anode itself giving a current pulse in the external circuit. It's now time to detail a few points that have obviously been 'glossed over' until now, namely the following:

1 Why do we only have to consider one domain at a time? Why don't several domains travel along the device at the same time?
2 Why should there be any charge fluctuation in the first place to form a domain?
3 What happens if the device is too short for a domain to fully form before reaching the anode?

Starting with (1), only one domain forms at a time because virtually all the field across the device is contained within the high-field domain region. Therefore the field throughout the rest of the device is reduced below the threshold field when a domain forms, and further domains *cannot develop*. The first domain to form 'hogs' the field until it reaches the anode and is lost. Another new domain can then develop. For point (2), a charge fluctuation can be expected to develop wherever we have a *doping fluctuation* in the device. A doping fluctuation is likely to occur near the metallic cathode contact where we change from very high electron concentrations in the metal contact itself, to much lower electron concentrations in the semiconductor. However, nowadays we deliberately fabricate the TED as shown in Figure 5.5 so that domains always form in the same high-field region of the device. Point (3) leads to the possibility of a different mode of operation of the TED. If the domain velocity is v_d and the active device length is L, then the time for the domain to travel to the anode, τ_{DOMAIN} is given by

$$\tau_{DOMAIN} = \frac{L}{v_d} \tag{5.31}$$

However, it takes time τ_d (given by equation (5.2)) for the domain to form. So, if the domain reaches the anode in a time less than τ_d, the domain will not have completely formed (before being lost to the external circuit). The conditions imposed by this situation lead to two classes of TED.

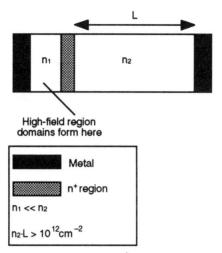

Figure 5.5 TED structure.

Calculation

Find the device parameters that define whether a domain completely forms, or not, within a TED assuming the effective drift field acting on the domain is 15 kV/cm.

A domain *will* completely form if the time for the domain to travel to the anode is greater than the time for the domain to form $\tau_{\text{DOMAIN}} > \tau_d$) or

$$\frac{L}{v_d} > \frac{\varepsilon_r \varepsilon_0}{q N_d \mu_d}$$

A simple rearrangement gives

$$N_d L > \frac{\varepsilon_0 \varepsilon_r v_d}{q \mu_d}$$

Substituting $v_d = \mu_d E_{\text{eff}}$ gives

$$N_d L > \frac{\varepsilon_0 \varepsilon_r E_{\text{eff}}}{q}$$

Hence,

$$N_d L > 1.1 \times 10^{11} \, \text{cm}^{-2}$$

A more accurate calculation will yield

$$N_d L > 10^{12} \, \text{cm}^{-2}$$

We therefore see that the *device criterion* of doping multiplied by device length $= 10^{12} \, \text{cm}^{-2}$ separates two classes of TED, one in which a stable domain is formed and another in which the domain *grows* as it travels to the anode. Devices of the latter type may be used in microwave amplifier circuits.

In *real* electronic circuits the frequency of operation is clearly dependent on the domain transit time (i.e. the length of the device) *and* the external resonant circuit of which the device is a part. TEDs may also be operated either pulsed or CW (CW or *continuous wave* means the same thing as *continuously* in this respect) and may give microwave power of the order of a watt at frequencies in excess of 10 GHz. For these reasons the TED is almost always found to be *the device* used in low-power microwave applications.

p–n junction

I must start this chapter with a word of warning. The p–n junction is often described as the simplest semiconductor device with which to start learning electronics. I won't quibble over the fact that *simpler* device structures exist (photoconductors, for example) since, *structurally* speaking, the p–n junction *is* quite simple. However, be warned. Understanding how the p–n junction actually works physically is in fact quite difficult! For those of you who have done a little electronics before, and are now using this book for your further education studies, who think you know how a p–n junction works, I can assure you, you don't. This may actually be the toughest chapter in the book, tougher even than the quantum mechanics section. It is arguably the most important *device* chapter in the book since the humble p–n junction takes us on to an understanding of the LED, the photodiode, the LASER and, of course, the bipolar transistor. This is therefore a fundamentally important chapter for an understanding of electronic devices.

6.1   The p–n junction in thermal equilibrium

The p–n junction is *the* basic element of all bipolar devices, hence its importance. Its main electrical property is that it rectifies (allows current to flow easily in one direction only). The p–n junction is often just called a diode, but we saw the confusion this might cause in the last chapter since *diode* just means a two-terminal device. However, during this chapter I shall frequently use the term 'diode', meaning the p–n junction. There are many types of p–n junction diode used for many different applications, such as the following:

(a) photodiode – light sensitive diode,
(b) LED – light emitting diode,
(c) varactor diode – variable capacitance diode,

to name just a few. As we progress we shall see how all these devices work, but for the moment we'll stick with the basic p–n junction. This is the first time we consider a device utilizing *both carrier types* so we'll take things slowly at first. We already know what the energy-level diagrams look like for *either* a p- or n-type semiconductor. They're shown again in Figure 6.1. What we don't know is what happens when we form a p–n junction as in Figure 6.2. I need to dwell for just a moment on the formation of this junction. In Figure 6.2 we *haven't* simply *pushed* a piece of p-type silicon into close contact with a piece of n-type silicon. Unfortunately that approach will not work. There are some simple reasons, and some not so simple reasons why pushing two bits of silicon together in this way will not work. Some simple reasons are the following:

1 There will only be very few points of contact and any current flow would be restricted to these few points instead of the whole surface area of the junction.
2 Silicon that has been exposed to the air always has a thin oxide coating on its surface called the 'native oxide'. This oxide, although thin, is a very good insulator and will prevent current flow.

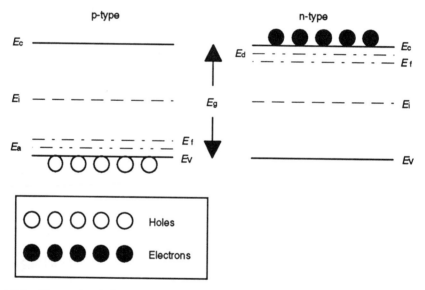

Figure 6.1   Energy-level diagrams.

A not so simple reason involves the bonding arrangements at the silicon surface. If we forget about the native oxide for a moment, and just consider the single-crystal silicon surface, then it is clear that something funny must happen at the surface. Why? Because in the bulk of the material there is no problem in satisfying all the silicon covalent bonds; there are plenty of near neighbours to do this. But the surface silicon atoms don't have any neighbours *above* them to bond with. This leads to odd types of bond reconstruction going on at the surface, as well as the possibility of *dangling bonds*, which are unsatisfied bonds just sitting there. In practice, unsatisfied bonds are likely to grab some contaminant from the

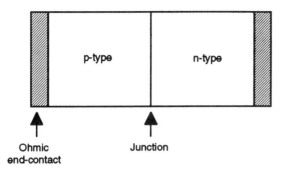

Figure 6.2   p–n junction.

air but you can still run into dangling bond trouble even when growing silicon dioxide on clean silicon surfaces. This particular problem led to considerable technological problems in early field-effect transistor work (and still can today) where we have an oxide layer on silicon as part of the device structure. At the interface between the oxide and the single-crystal silicon, those dangling bonds and other *surface states* can cause us a lot of trouble.

Returning to our p–n junction formed by pushing two pieces of silicon together, these surface states would cause us enormous problems in this instance as well. Basically, you wouldn't get a very good-looking rectifying contact by this approach. I'm spending a little time on this to show you why the diode takes the form it does today. So, how do we overcome this surface problem? We form the p–n junction in *the bulk* of the semiconductor, away from the surface as much as possible. Figure 6.3 shows a more realistic diode structure. Note that a portion of the p–n junction is still at the surface, around the periphery of the p$^+$ region (it has to be because of the way the diode is made), but this region is small compared to the rest of the junction which lies in the bulk of the silicon. In addition, a good-quality thermally grown oxide covers the surface junction region, which helps to minimize problems here. This oxide, if grown properly, *passivates* the silicon surface, reducing the number of surface states as well as protecting the silicon surface from the ingress of contaminants. I don't want to get into the technology of how devices are fabricated as that requires a book in itself, and there are some very good ones available. However, a couple of steps used in fabricating the diode of Figure 6.3 should be covered. The p$^+$ (highly doped p) region is formed by either *diffusing* boron into the silicon surface at high temperature, or by *ion-implanting* boron into the silicon. Remember, all we need is for $N_A > N_D$ in order to convert the n-type *substrate* (starting single-crystal silicon slice in which we make the device) to p-type. The p$^+$ region is confined by ion-implanting or diffusing the boron *through a hole* etched through the top insulating oxide. For obvious reasons this hole is called a *window*. The hole itself is etched either using

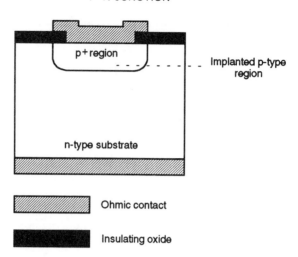

Figure 6.3 Fabricated diode.

hydrofluoric acid (HF) or an *ion-etcher*. The patterns which define where the oxide may be etched are put down *photolithographically*. That's as much as I want to say about the technology. I now want to get back to the question of the energy-level diagram for the p–n junction.

Going back to Figure 6.1 we could push the two energy-level diagrams together, as shown in Figure 6.4, to get a combined energy-level diagram for a p–n junction. Although we can't make a diode by pushing together the two bits of doped silicon, that doesn't stop us pushing the two diagrams together to see what happens. Look at the result in Figure 6.4. There is a big discontinuity in the Fermi-level across the p–n junction. Is this reasonable? Can this be physically correct? In fact, Figure 6.4 is *incorrect* and we have enough background knowledge to find out why this is so.

Figure 6.5 shows an idealized p–n junction. A piece of p-type silicon has formed an ideal junction with a piece of n-type material. The solid line drawn separating the two differently doped materials is called the *metallurgical junction*. Something has got to 'give' here. We have got silicon with lots of holes (and therefore very few electrons, remember $np = n_i^2$) in contact with silicon containing lots of electrons. Not only do the holes and electrons want to recombine but we also have ideal conditions for carrier diffusion. Lots of holes on the right-hand side of the junction want to diffuse to the left and lots of electrons on the left-hand side of the junction want to move to the right. And this is what they do! Now we come to the only tricky bit with this simple picture. We need to differentiate between the carriers (the electrons and holes) and the *donors and acceptors* that give rise to them. Remember the donors and acceptors are elements (such as arsenic and boron) that are incorporated into the silicon lattice and are therefore *fixed in position*. The donors and acceptors *do not move* (unless

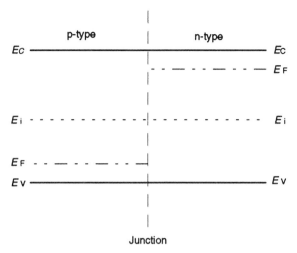

Figure 6.4 Incorrect band diagram.

you heat up the semiconductor so they can *diffuse*!). The electrons and holes that come from them are, however, *free* to move. I think you can see that as holes diffuse to the left of the metallurgical junction and combine with the electrons on that side, they leave behind negatively charged acceptor centres (recall the definitions of donors and acceptors). Similarly, electrons diffusing to the right will leave behind positively charged donor centres. Clearly, this diffusion process cannot go on forever. The increasing amount of fixed charge that is being uncovered wants to electrostatically attract the carriers that are trying to diffuse away (the acceptor centres want to keep the holes, the donor centres want to keep the electrons). Another way of looking at this, which is why we covered it in the last chapter, is that the *electric field* produced by the *fixed charges* builds up to a value that slows down the diffusion process. The diffusion process doesn't come to a dead halt, but it does reach an *equilibrium condition*. Figure 6.6 shows the equilibrium condition that is reached. The region between the two dotted lines is the region the free carriers have left, leaving behind the uncovered charge or *space-charge*. This region is therefore known as the space-charge region or the *depletion* region, since it is depleted of free carriers. Similar regions occur in other devices (such as MOSFETS and bipolar transistors) and are fortunately given the same name. Away from this junction region (in which charge movement has occurred) we enter material where the carriers are still sitting on the dopant centres, i.e. *neutral* material. Note the direction of the electric field due to the uncovered dopant charge and prove to yourself that this hinders any further free carrier diffusion.

We are now in a position to derive some equations. In this equilibrium situation there is *no net current flow*. There is still *some* current flow, for instance any *minority carriers* that find themselves near the junction region will not be

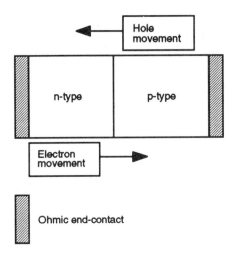

Figure 6.5   Idealized p–n junction.

hindered by the electric field at the junction. On the contrary, they will be accelerated across the junction to the other side. Under equilibrium conditions this current flow will be balanced by an opposing diffusion current. The important thing to remember is that, as stated earlier, things haven't come to a dead halt, currents are flowing all the time, but the *net* current flow *is zero*. Let's get on with the equations!

If the net hole current is zero (as is the net electron current), then

$$J_p = J_p(DRIFT) + J_p(DIFFUSION) = 0 \qquad (6.1)$$

where

J_p is the hole current density ($A\,cm^{-2}$),
the drift current is due to the electric field at the junction,
the diffusion current is due to the concentration gradient.

Here's another first for us. This device has drift *and* diffusion currents going on at the same time. We only needed to consider drift currents in the Gunn diode. In order to turn (6.1) into a mathematical equation we now have to start digging out some of the Chapter 4 material. For the total hole current density (drift plus diffusion) we need equation (4.60), which gives us

$$J_p = q\mu_p p E_x - qD_p \frac{dp}{dx} = 0 \qquad (4.60)$$

Substitute the electric field in the x-direction with the slope of the intrinsic Fermi-level according to equation (4.31), and use the Einstein relation (equation (4.58)) to substitute for the hole diffusion coefficient. Making these substitutions we get,

$$J_p = \mu_p \left(p \frac{dE_i}{dx} - kT \frac{dp}{dx} \right) = 0 \qquad (6.2)$$

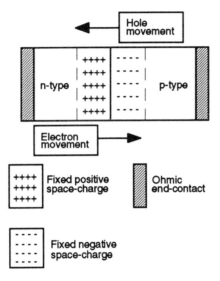

Figure 6.6   Idealized p–n junction.

We now want to substitute for the dp/dx. To do this go back to equation (3.12):

$$p = n_i \exp\left(\frac{E_i - E_F}{kT}\right) \tag{3.12}$$

Differentiate equation (3.12) with respect to x. Remember that to differentiate an equation like this that has an exponential function in it, first take log to base e *and then* differentiate the equation. You should get

$$\frac{dp}{dx} = \frac{p}{kT}\left(\frac{dE_i}{dx} - \frac{dE_F}{dx}\right) \tag{6.3}$$

Substituting this result into equation (6.2) gives us

$$J_p = \mu_p p \frac{dE_F}{dx} = 0 \tag{6.4}$$

from which we conclude that

$$\frac{dE_F}{dx} = 0 \tag{6.5}$$

which states that the Fermi-level is a constant (a horizontal straight line on the energy-level diagram) with distance through the diode.

We can go through *exactly* the same process for the net electron current density (do this for yourself) to obtain

$$J_n = \mu_n n \frac{dE_F}{dx} = 0 \tag{6.6}$$

which again gives us equation (6.5) as a result. We conclude that for *zero net electron and hole currents* across the depletion region in the p–n junction, *the Fermi-level is a constant*. To put it another way, since we are considering the p–n junction in thermal equilibrium, a thermal equilibrium condition is that the Fermi-level must remain constant. This is why Figure 6.4 is incorrect. Although we have shown that the Fermi-level is horizontal for electrons and holes separately, we haven't shown that there is only *one* Fermi-level to consider. If we go into the neutral n-type semiconductor, well away from the junction region, *we know* that only one Fermi-level is required and the situation is as shown in Figure 3.11. The same thing applies in the bulk of the neutral p-type semiconductor. Again, one Fermi-level is sufficient to describe the physical situation. But we have just shown that the Fermi-levels for electrons and holes are constant throughout the whole p–n junction in thermal equilibrium. Therefore, since there is only *one* Fermi-level in the neutral regions, there can only be one Fermi-level through the depleted junction region as well. That may have been a long-winded argument to some of you who intuitively felt that there *was* only one level to consider, but for the mathematicians amongst you we only *proved* that the Fermi-level was constant for electrons and holes, not that the Fermi-level also had the same value for both! After all that, we can now draw the energy-level diagram for the p–n junction in thermal equilibrium as in Figure 6.7. You can now see how to draw this type of diagram for yourself, and indeed more complicated diagrams with

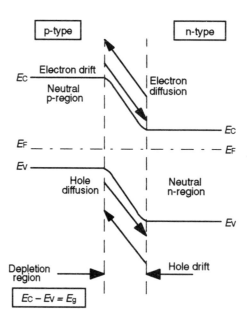

Figure 6.7   p–n junction energy diagram.

several different p- and n-layers. You bring together the separate energy-level diagrams for p- and n-type material, but you don't align the conduction and valence bands as we did in Figure 6.4. Instead you *align the Fermi-levels*. Leave a little gap between the two separate diagrams so you can connect the displaced conduction and valence band edges together smoothly. Finally, as you link up the band edges, make sure their separation is kept constant at the band-gap energy. That's all there is to making simple energy-level diagrams for real devices. In fact, right now you could have a go at drawing the energy-level diagram for an npn transistor. Try putting together the energy-level diagrams for a high n-type doped layer and a low n-type doped layer with a low p-type doped layer sandwiched in between. When you've tried that take a look at Figure 8.9 in Chapter 8 to see how you've done.

Returning to Figure 6.7, note the opposing drift and diffusion currents. Remember that these currents are flowing all the time but that in thermal equilibrium their effects cancel out to give us zero net current flow. Under non-thermal equilibrium conditions (i.e. when we finally get round to putting some volts on this thing) one of the current flow mechanisms is going to dominate over the other to give us a non-zero net current flow, at least in one direction. Why should there be some directionality in the current flow through this device at all? Look carefully around the depletion region of the diode. The electrons that want to *diffuse* from the n-type layer (lots of electrons) to the p-layer (not many electrons) have a *potential barrier* in the way formed by that slope in the conduction band. The very few thermally generated electrons in the p-layer that find themselves near the junction region, however, don't have any such *barrier* to consider. In fact they want to slide down the slope; remember under drift conditions electrons *sink*. Now look at what is happening in the valence band. The holes that want to diffuse from the p-type layer to the n-type layer also have a potential barrier in the way (recall from energy considerations that the hole barrier *must* go in the opposite direction to an electron barrier). The few minority holes produced near the n-side of the junction, however, are helped on their way from the n-layer into the p-layer by the electric field produced by the uncovered charges in the depletion region. We shall look at the electric fields and potential produced by a p–n junction in a moment, but just before we do that, I'd like to calculate the height of that potential barrier in the diode.

6.2 p–n junction barrier height

The potential barrier height we're trying to calculate is designated V_{bi} in Figure 6.8. This potential barrier height across a p–n junction is known as the *built-in potential* and also as the *junction potential*. The potential *energy* that this potential barrier corresponds to is given by equation (4.34) and is qV_{bi}. Remember that electron *energy* is positive upwards in these energy-level diagrams, so from equation (4.34) electron *potentials* are going to be measured positive *downwards*.

To make things even more complicated, the *hole* potentials and energies are of course positive in the opposite direction to the electrons. You can see that trying to keep track of all these plus and minus signs can quickly become a nightmare. The cop-out solution, which I am going to take, is to just consider *magnitudes*. The magnitude of qV_{bi} is seen to be the sum of the magnitudes of qV_p and qV_n. Considering the neutral p-region, the energy qV_p is just the difference in energy between the Fermi-level and the intrinsic Fermi-level on the p-side of the junction:

$$qV_p = -(E_i - E_F) \tag{6.7}$$

Now you can see why we chose E_i in equation (4.31). The intrinsic Fermi-level is a very useful *reference level* in a semiconductor.

Using equation (3.12) we can now substitute for $E_i - E_F$ to obtain

$$|V_p| = \frac{kT}{q} \ln\frac{N_A}{n_i} \tag{6.8}$$

where we assume that we have full ionization of the acceptor level and so $p = N_A$. In exactly the same way we can now consider the neutral n-region and find the magnitude of V_n:

$$qV_n = -(E_i - E_F) \tag{6.9}$$

Substitute for $E_i - E_F$ using equation (3.9):

$$|V_n| = \frac{kT}{q} \ln\frac{N_D}{n_i} \tag{6.10}$$

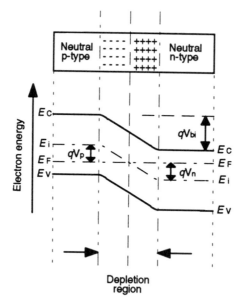

Figure 6.8 p–n junction in thermal equilibrium.

where again we have assumed full ionization, this time of the donor level, and so $n = N_D$. The built-in voltage is now going to be given by the sum of V_n and V_p hence,

$$| V_{bi} | = | V_n | + | V_p | = \frac{kT}{q} \ln \frac{N_A N_D}{n^2_i} \qquad (6.11)$$

Calculation

Let's quickly calculate the built-in potential of a p–n junction doped with $10^{18} \, cm^{-3}$ acceptors on the p-side and $10^{15} \, cm^{-3}$ donors on the n-side. Simply putting the values into equation (6.11) we get

$$V_{bi} = (0.0259) \ln(10^{18} \times 10^{15})/(1.4 \times 10^{10})^2 = 0.76 \, V$$

Notice something pretty important here. In some elementary texts you may see that there is a forward voltage drop of 0.6 volt before the p–n junction starts to conduct properly. The statement itself is incorrect, as we shall see later, but the voltage drop that is being talked about is basically the built-in voltage. We have just seen that we've got a built-in voltage quite different from 0.6 volt, and in general of course you will get a different voltage because *it depends on the doping in the p- and n-regions*. There is no *constant* voltage drop for *all* p–n junctions; it depends on the doping!

We've done quite a bit of work on the p–n junction so far and we still haven't put any volts on it yet! There's just a little bit more work to cover before we apply (voltage) bias to the junction. We want to see the form of the electric field and potential across the junction.

6.3   Depletion approximation, electric field and potential

We're still not applying bias to the junction, so the physical situation shown in Figure 6.9 is for a p–n junction in thermal equilibrium. The only charge we need to consider is the uncovered dopant charge in the depletion region; the rest of the p–n junction is neutral. We know from our work on the Gunn diode that from a charge distribution we can quickly draw the electric field and potential distributions without doing any maths (see Section 5.3). The electric field and potential distributions across the depletion region are shown in Figure 6.9, which requires some explanation. Let's start with the charge distribution. I've got p-type material on the left of the metallurgical junction and n-type material on the right. This means the uncovered dopant charge will be negative on the left of the junction and positive on the right, as shown. The charge *distribution*, however, will *not* be nice and rectangular as shown in the diagram. In reality the uncovered charge will 'tail off' into the neutral regions and the edge will not be sharply

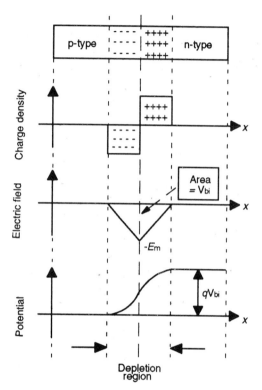

Figure 6.9   p–n junction at zero bias.

defined as in the diagram. In practice this *transition region* where the charge gradually drops off into the neutral region is usually very short, which means that our *depletion approximation*, where we have picked an abrupt transition from the charge region to the neutral region, is in fact quite a good approximation.

We're now ready to start doing the mental integrations to get the field and the potential. Starting at the left-hand edge of the negative charge distribution and integrating to the right we will produce an electric field that starts at zero and slowly increases in a *negative* (the field takes the same sign as the charge) direction. We then reach the metallurgical junction. At this point the charge takes on a change of sign, so further integration leads to a *decrease* in the magnitude of the electric field (it increases positively), as shown. Clearly, the electric-field distribution falls to zero at the edge of the charge distribution as there will be no electric field without charge. We are now ready to mentally integrate the electric-field distribution to get the potential diagram. The first thing to remember is that the sign of the electric field is opposite to that of the potential so integrating the *negative* field will lead to a *positive* potential. Since the electric field is negative everywhere, then (from our integration) the potential will be positive everywhere, but you can define the *zero* of potential wherever you like. I'll have to come back to this point in a moment. Starting at the left-hand edge of the depletion region we start integrating. As we integrate towards the metallurgical junction the potential will slowly increase. On reaching the junction there will be *no sign change* (the electric field is negative throughout) so the potential will continue to increase. There is, however, a *change* of slope in the electric field at the metallurgical junction which feeds through into a change in the rate of increase in the potential at this point. Since the change in potential in going across the junction is going to be the built-in voltage V_{bi}, it follows that the *area under the triangle* in the electric field distribution must equal V_{bi} as well.

Although we've painlessly got the field and potential distributions without doing any maths (and we could get other quantities, too, if we drew to scale on graph paper and took measurements!), unfortunately we're going to have to resort to the maths in a moment to fully appreciate what's going on. Before doing so, just a couple of points. The junction we're looking at here is called an *abrupt junction* because the doping abruptly changes from p- to n-type at the metallurgical junction. The doping change need not be abrupt; a smooth change in doping density would lead to a *graded* junction which has many applications, especially in bipolar transistors. An abrupt junction that has one side *much more* highly doped than the other is called a *one-sided abrupt junction* (stands to reason I suppose).

Before moving on to the maths, I want to redraw Figure 6.9 with the n-side on the left and the p-side on the right. You might feel that this is a bit of a waste of time since everything should just come out the same, which of course it should. But does it? Look at Figure 6.10. The electric field is now a positive quantity!! Has something gone wrong?? It needs a bit of thought this one. In Figure 6.9 the negative electric field means that since the field is pointing towards negative x it will tend to push holes towards the left and electrons towards the right. In other words it wants to keep the free carriers in their respective regions. We know this is right since equilibrium comes about by the electric field produced, preventing further movement of carriers, so Figure 6.9 is OK. One further point about the electric-field distribution diagram. The *value* on the y-axis denotes the *magnitude* of the electric field. The *sign* on the y-axis denotes the electric-field direction (plus or minus x). I find this confusing myself and need to think about it every time. You should now see that this makes Figure 6.10 quite OK. The electric field is now a positive quantity, the force acting on the electrons will be to the left (keeping them in the n-region) and the force on the holes will be to the right (keeping them in the p-region). The only other thing that needs some explanation now is the potential diagram. For this potential distribution I have integrated the electric field and defined the zero potential at $x = x_p$. If I had integrated from left to right in the normal way I would have got the dotted curve as shown. Physically, it doesn't matter either way because all that matters are *differences* in potential not the absolute value of the potential itself, and in both cases crossing the junction means a potential (change) difference of V_{bi}! If you look at some of the

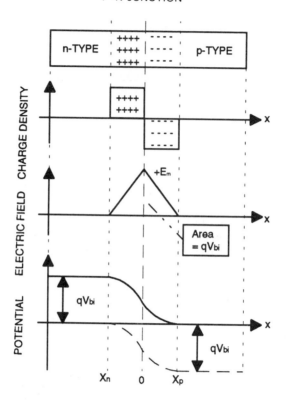

Figure 6.10 p–n junction at zero bias.

other standard texts you will see that some people make the potential *zero at the metallurgical junction*. We shall do this as well when we integrate the field equations but it isn't strictly necessary. You can *define* the *zero* of potential *wherever you want*, but I hope you see that it is actually unimportant since all that really matters are potential differences.

6.4 Mathematical formulation

We shall start by using Figure 6.11 to see what's going on. Overall, the space-charge region has got to be neutral. This should be clear since the uncovered charges will have been formed by electrons recombining with holes, one electron recombining with one hole. So in Figure 6.11 the amount of uncovered positive charge on the n-side of the metallurgical junction must be equal to the amount of uncovered negative charge on the p-side of the junction. The *widths* of the space-charge regions are drawn unequal because I am assuming that the *doping densities* on either side of the junction are unequal. Mathematically, we can write the overall space-charge neutrality condition as

$$N_A x_p = N_D x_n \tag{6.12}$$

This simple equation is already very useful. It tells us that the *higher* doped side of the junction has the *narrower* depletion width. The *total* depletion-layer width W will just be the sum of the individual depletion widths:

$$W = x_p + x_n \tag{6.13}$$

If we take the zero of the x-axis at the metallurgical junction, then Poisson's equation (equation (5.7)) will give us

$$\frac{d^2V}{dx^2} = +\frac{qN_A}{\varepsilon_{Si}} \text{ for } -x_p \le x < 0 \tag{6.14}$$

and for the n-side:

$$\frac{d^2V}{dx^2} = -\frac{qN_D}{\varepsilon_{Si}} \text{ for } 0 < x \le x_n \tag{6.15}$$

where

$$\varepsilon_{Si} = \varepsilon_0 \varepsilon_r$$

and

$$\varepsilon_r = \text{relative permittivity of silicon} = 11.9$$

Note something a little tricky in equations (6.14) and (6.15). You need to be *very*

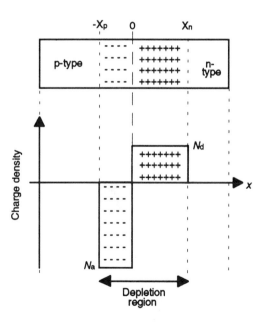

Figure 6.11 Space-charge neutrality.

careful with the signs now (we can't get away with magnitudes). Poisson's equation relates the second derivative of the potential to the *negative* of the charge density. The charge density due to the uncovered acceptors N_A is *negative* leading to the *plus* sign in equation (6.14). The charge density due to the uncovered donors is *positive* leading to the *minus* sign in equation (6.15). In order to get the equations for the electric field as a function of distance, we need to integrate equations (6.14) and (6.15). Although the actual integrations are straightforward, the application of the appropriate boundary condition isn't. For this reason I shall go through the integration of the equations fully. Starting with (6.14):

$$\frac{d^2V}{dx^2} = +\frac{qN_A}{\varepsilon_{Si}} \text{for } -x_p \le x < 0$$

Now integrate this expression with respect to x and don't forget the constant of integration,

$$E_p(x) = -\frac{dV}{dx} = -\frac{qN_A x}{\varepsilon_{Si}} + C_1 \tag{6.16}$$

where

E_p is the electric field on the p-side of the junction,
C_1 is the constant of integration.

To find the constant of integration we note that $E_p(x) = 0$ when $x = -x_p$. This leads to

$$C_1 = -\frac{qN_A x_p}{\varepsilon_{Si}}$$

Substituting for C_1 in equation (6.16) gives

$$E_p(x) = -\frac{qN_A(x + x_p)}{\varepsilon_{Si}} \tag{6.17}$$

We go through exactly the same procedure for the n-side of the junction starting with equation (6.15):

$$\frac{d^2V}{dx^2} = -\frac{qN_D}{\varepsilon_{Si}} \text{ for } 0 < x \le x_n$$

Integrate with respect to x:

$$E_n(x) = -\frac{dV}{dx} = \frac{qN_D x}{\varepsilon_{Si}} + C_2 \tag{6.18}$$

The boundary conditions give $E_n(x) = 0$ when $x = x_n$, which gives

$$C_2 = -\frac{qN_D x_n}{\varepsilon_{Si}}$$

Substituting the constant of integration into equation (6.18) leads to

$$E_n(x) = \frac{qN_D(x - x_n)}{\varepsilon_{Si}} \tag{6.19}$$

We thus have equations for the electric field through the depletion region of a p–n junction diode. We know from our pictorial integrations that the maximum electric field occurs at the metallurgical junction ($x = 0$) so if we put $x = 0$ into the above equations we shall obtain a *value* for the maximum electric field:

$$\mathbf{E}_m = -\frac{qN_D x_n}{\varepsilon_{Si}} = -\frac{qN_A x_p}{\varepsilon_{Si}} \tag{6.20}$$

Integrating the electric-field equations, of course, leads to the potential as a function of distance. *These integrations are simpler because the constant of integration is zero if we* choose the potential to be zero at $x = 0$. Hence, integrating the electric-field equations we obtain

$$V_p(x) = \frac{qN_A x^2}{2\varepsilon_{Si}} + \frac{qN_A x x_p}{\varepsilon_{Si}} \tag{6.21}$$

For $x \leq -x_p$ the potential is *constant* and equal to the value of the potential *at* $x = -x_p$. The value of the potential at $x = -x_p$ is

$$V_p(-x_p) = -\frac{qN_A x_p^2}{2\varepsilon_{Si}} \tag{6.22}$$

By exactly the same reasoning we get the potential on the n-side of the junction to be

$$V_n(x) = -\frac{qN_D x^2}{2\varepsilon_{Si}} + \frac{qN_D x_n x}{\varepsilon_{Si}} \tag{6.23}$$

And the potential at $x = x_n$ is

$$V_n(x_n) = \frac{qN_D x_n^2}{2\varepsilon_{Si}} \tag{6.24}$$

We know the total potential difference across the whole junction is just the built-in voltage, so

$$V_{bi} = V_n - V_p = \frac{q}{2\varepsilon_{Si}}(N_A x_p^2 + N_D x_n^2) \tag{6.25}$$

I didn't cheat this time by just considering magnitudes. The correct signs have been used. We can also write the built-in potential in terms of the maximum electric field and the depletion-layer width. Referring back to Figures 6.9 or 6.10 or 6.11, we know from the calculus that the built-in voltage is just the triangular area under the electric field curve. Basic geometry gives the area of a triangle as half the base length times the height; the base length is W and the height is $\mathbf{E}_m$, so the built-in voltage is

$$V_{bi} = \left| \frac{\mathbf{E}_m W}{2} \right| \tag{6.26}$$

We now want an expression for the total depletion-layer width, W. We therefore need an expression for $x_p + x_n$. Since $N_A x_p = N_D x_n$ we can rewrite the built-in potential equation (6.25) as

$$V_{bi} = \frac{q}{2\varepsilon_{Si}} N_A^2 x_p^2 \left(\frac{N_A + N_D}{N_A N_D} \right) = \frac{q}{2\varepsilon_{Si}} N_D^2 x_n^2 \left(\frac{N_A + N_D}{N_A N_D} \right) \tag{6.27}$$

From equation (6.27) we can extract the depletion-layer widths for the p- and n-sides of the junction:

$$x_p = \frac{1}{N_A} \sqrt{\frac{2\varepsilon_{Si} V_{bi} N_A N_D}{q(N_A + N_D)}}$$ (6.28)

and for the n-side of the junction:

$$x_n = \frac{1}{N_D} \sqrt{\frac{2\varepsilon_{Si} V_{bi} N_A N_D}{q(N_A + N_D)}}$$ (6.29)

The total depletion-layer width is the sum of equations (6.28) and (6.29):

$$W = \sqrt{\frac{2\varepsilon_{Si}(N_A + N_D)V_{bi}}{q N_A N_D}}$$ (6.30)

Calculation

Calculate the built-in potential and the total depletion-layer width for an abrupt p–n junction diode with p-doping of $10^{17}\,cm^{-3}$ and n-doping of $10^{15}\,cm^{-3}$.

For the built-in potential use equation (6.11) and substitute the appropriate values:

$$V_{bi} = (0.0259)\ln(10^{17} \times 10^{15})/(1.4 \times 10^{10})^2 = 0.7\,V$$

For the total depletion-layer width use the equation we have just derived, equation (6.30):

$W = ((2 \times 8.854 \times 10^{-14} \times 11.9 \times 1.01 \times 10^{17} \times 0.7)/(1.6 \times 10^{-19} \times 10^{32}))^{0.5}$
$W = 9.65 \times 10^{-5}\,cm = 0.965\,\mu m.$

So the depletion-layer widths we're talking about for typical diode doping levels are about a micron, where a micron is $10^{-6}\,m$, also written μm. You should always have a 'feel' for typical device (and material) dimensions and parameters. Do you remember the rough order of magnitude for the electron relaxation time?

6.5   One-sided, abrupt p–n junction

As mentioned earlier, when the dopant concentration on one side of an abrupt junction is much higher than the other side, the junction is called a one-sided abrupt junction. Figure 6.12 shows the charge, electric field, and potential distributions as a function of distance for the one-sided abrupt junction. By now these 'mental' integrations should be becoming straightforward. In this set of diagrams all that needs mention is the shape of the electric field and the potential distribution. Because the acceptor doping is very much greater than the donor doping, virtually all of the space-charge will be on the donor, n-side, of the junction. In reality, there will be a very small depletion region extending into the p-side of the junction, but we usually omit this when drawing the diagrams. Hence in Figure 6.12 I don't show the electric field extending a little way into the p-side of the junction (which it of course does) and I similarly omit the tiny bit of potential which also lies on the p-side. These simplifications are normally employed when discussing this device. We are considering this type of diode because it is often fabricated in practice. It is after all quite simple to heavily overdope a piece of p- or n-type silicon to form this structure. We have seen all the mathematical formulation before in Section 6.4 and we shall use these results to obtain the simpler formulae for the one-sided junction.

For the width of the depletion region we only need to consider the width of the space-charge region on the n-side of the junction. Hence we start with equation (6.29):

$$x_n = \frac{1}{N_D} \sqrt{\frac{2\varepsilon_{Si} V_{bi} N_A N_D}{q(N_A + N_D)}} \tag{6.29}$$

But we also have $N_A \gg N_D$ so the equation simplifies to

$$W = x_n = \sqrt{\frac{2\varepsilon_{Si} V_{bi}}{q N_D}} \quad \text{ABRUPT ONE-SIDED JUNCTION} \tag{6.30}$$

The electric-field distribution is as given in equation (6.18):

$$E_n(x) = \frac{q N_D x}{\varepsilon_{Si}} + C_2 \tag{6.18}$$

In this case the boundary condition is $E_n(x) = -E_m$ when $x = 0$. This boundary condition gives us C_2 and equation (6.18) becomes, for the one-sided junction,

$$E(x) = -E_m + \frac{q N_D x}{\varepsilon_{Si}} \tag{6.31}$$

Since the second boundary condition is that $E(x) = 0$ at $x = W$ we can write for E_m,

$$E_m = \frac{q N_D W}{\varepsilon_{Si}} \tag{6.32}$$

This equation for the maximum electric field can now be substituted into equation (6.31) to give two more equivalent equations for the electric-field distribution:

$$E(x) = \frac{q N_D}{\varepsilon_{Si}}(x - W) \tag{6.33}$$

and

$$E(x) = -E_m\left(1 - \frac{x}{W}\right) \tag{6.34}$$

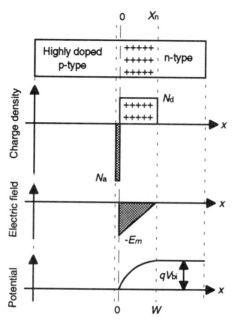

Figure 6.12　One-sided abrupt junction.

And, as before, the potential is found by integrating the electric field with respect to x:

$$V(x) = -\int_0^x E(x)\,dx = E_m\left(x - \frac{x^2}{2W}\right) + const \qquad (6.35)$$

The boundary condition in this case is that the potential is zero at $x = 0$, which means the constant of integration is also zero. Using the results of this section means that we can write the potential distribution as

$$V(x) = \frac{V_{bi}x}{W}\left(2 - \frac{x}{W}\right) \quad \text{ABRUPT ONE-SIDED JUNCTION} \qquad (6.36)$$

We've spent quite long enough now looking at the p–n junction diode, just sitting there, with no applied bias. It's time to see what this thing does when we apply some volts!

6.6 Applying bias to the p–n junction

Clearly, there are two ways of applying a voltage (bias) to the p–n junction. We can either connect the positive terminal of the voltage source to the n-side or the p-side of the junction. The first thing we need to see is how current flows through the diode under these two conditions.

Figure 6.13 is a *current–voltage* or *I–V* characteristic of a p–n junction. From the diagram it is seen that current flows easily in one direction (forward bias) and that there is very little current flow with the bias reversed (reverse bias). See how quickly the current increases in the forward direction with increasing voltage; the increase is in fact exponential. Also note the constant (very low) current that flows under reverse bias, until breakdown of the junction occurs. Another key feature of this ideal p–n junction *I–V* characteristic is that there is always a current flowing if there is an applied bias (even in reverse bias). *There is no turn-on voltage.* We saw from the last section that there is no constant turn-on voltage, the barrier height was a function of the p- and n-type doping levels. We now see that there is in fact no turn-on voltage at all. The diode always conducts (though much more heavily in the forward direction) providing there is bias. What you can do, however, and this *is* often done in practice, is to *define* a turn-on voltage as the forward bias required to produce a given amount of forward current. For instance, if you are designing a circuit and you need at least 1 mA flowing through a diode (for the circuit to work), and this in turn means that the diode needs say 0.7 volt forward bias, then from a circuit design point of view you *could* say the diode has a 0.7 volt turn-on voltage. However, to my mind this is just playing around with words, and does not help us much with the physics. From our work

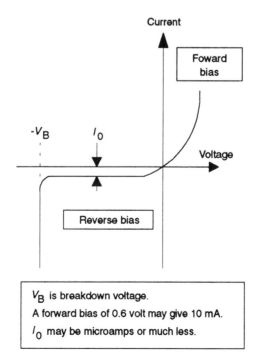

Figure 6.13   p–n junction *I–V* characteristic.

in the previous chapters we have covered enough material to understand the operation of the p–n junction from a physics point of view.

Figure 6.14 shows the operation of a p–n junction under thermal equilibrium conditions (no applied bias), and with forward and reverse biases applied. Also shown in the diagram are the energy band and potential diagrams associated with each bias case. At this stage we can qualitatively understand what is going on and a little later we'll cover the maths. Returning to the diagram, the equilibrium case should hold no surprises for you; the potential and energy-level diagrams have been seen before. The only thing to note is that the energy equivalent of a potential step of V_{bi} is given by qV_{bi} as we have already seen in equation (4.28). From Section 4.7 we also know that bands move when we apply an electric field to semiconductors, and that section tells us how the bands behave. We don't need to go to the maths to see what will happen; we should be able to work out what's going on from Figure 6.14. Let's take the *reverse bias* case first. The electric field we are applying to the p–n junction is in *the same direction* as the built-in electric field due to the uncovered charges in the depletion region. We can add electric fields (vectorially) which in this simple one-dimensional case means that we simply add them algebraically. This greater electric field we now have across the p–n junction must be provided by more uncovered charge in the depletion region, hence *under reverse bias conditions the depletion region will widen*. Let's look at this another way. The potential we're applying will want to draw more holes from the p-side into the n-side, and vice-versa, and this is precisely what happens. When does this transfer of free carriers stop? Just as in the equilibrium case, the charge transfer stops when the field at the junction builds up to a value sufficient to prevent further movement of carriers (remember, charge transfer *doesn't* come to a dead stop, but *net* charge transfer does). Now look at the energy-level and potential diagrams for the reverse-bias case. As we might expect the potential across the junction is greater than the thermal equilibrium case. Why? The uncovered depletion-region charge is greater, therefore the electric field at the junction is greater (integrate), therefore the potential difference across the junction is greater (integrate again)! It all hangs together so far.

If we now look at the energy-level diagram in the reverse bias situation we see that the height of the sloping part of the diagram, across the depletion region, is much bigger than the thermal equilibrium case, as we would expect from our work in Section 4.7. Note that both the potential and energy bands are *flat and horizontal* outside the depletion region. The reason for this is that we assume the neutral parts of the diode are sufficiently conducting, that *no volts are dropped* across these neutral regions. Consequently, we assume that *all the volts are dropped across the depletion region*. This assumption is generally valid but will of course fail in lightly doped (resistive) material, and when *very* large currents start to flow. Since the reverse biased p–n junction is holding off all the applied volts (virtually no current is flowing), then it is clear that the junction region has the applied voltage plus the built-in voltage across it, as shown in the diagram. The potential across the diode has therefore been increased from V_{bi} to $(V_{bi} + V_{APPLIED})$.

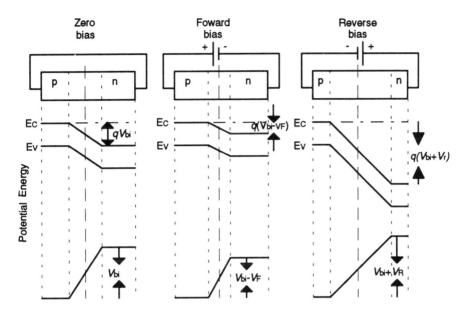

Figure 6.14 Biasing the p–n junction.

What are the consequences of this for current flow? Let's take the diffusion currents first. We now have a much greater barrier height for the electrons *and* holes to get over and as the current will *decrease* exponentially with barrier height *increase*, we can expect a very drastically reduced diffusion current (although it was already pretty small in equilibrium). The drift current that flows won't change that much. Why? Because it's still those minority carrier electrons and holes that are thermally generated within a diffusion length or so of the depletion-layer edge that will be swept across the depletion region by the strong electric field that exists in that region. So, those electrons produced close to the depletion edge *on the p-side of the junction* will be swept across, and the same goes for the holes produced on the n-side of the junction. That is, of course, a simple picture and we can easily think up situations where this wouldn't occur. For instance, if the increased depletion-layer width enclosed some carrier generation centres (defects, impurities) we would see an increase in the drift current with reverse bias. If in addition further bias increase caused more generation centres to enter the depletion region we would actually see an *increase* in the reverse current with increasing reverse bias. In the ideal case where we have a constant reverse drift current flowing, this current is called the *reverse saturation current*.

We now need to go through the same analysis for the p–n junction under *forward bias* conditions. Referring to Figure 6.14 you see that forward bias means that the applied electric field now *opposes* the internal built-in field. This being the case, and following exactly the same arguments as above, means that the depletion-layer width will be reduced under forward bias conditions. The reduction in depletion-layer width is accompanied by a reduction in the energy barrier heights for electrons and holes and by a reduction in the potential difference across the junction. That much should be clear from our discussion of reverse bias. What is less certain is what happens to the current flow. Under these forward bias conditions, again, the drift currents are practically unaltered since any minority carriers within a diffusion length of the depletion-layer edge are still swept across. The big change comes with the diffusion currents that can now flow! In forward bias we have *reduced* the potential barrier from V_{bi} to $(V_{bi} - V_{APPLIED})$ and we therefore now have very much greater diffusion currents flowing than in the equilibrium case. The argument for the *reduction* in barrier height with *forward bias* follows exactly the same lines as the barrier height *increase* resulting from the application of a *reverse bias*. This is why the *I–V* curve shows such a considerable (exponential) increase in current flow with applied forward bias.

Before trying to summarize what we have covered in this section I need to spell out something about this forward bias situation. It is a little simplistic to think of the barrier height being reduced by the applied voltage under these forward bias conditions. Why? Well, what if I apply a forward bias of 100 volts and the built-in bias is 0.6 volts? Do the bands bend the other way by 99.4 volts? What happens? This apparent problem arises because we are looking at the diode in isolation whereas in a real circuit it is likely to have a current limiting resistor in series with it. In reality we cannot simply put 100 volts forward bias across the diode, because with just a fraction of a volt forward bias we would practically flatten the bands out so that there would be virtually no limit to the hole and electron diffusion currents that would flow. This situation would quickly destroy the device by current heating. Figure 6.15 shows a more realistic practical situation. Under forward bias conditions the diode resistance is very small compared to the current limiting series resistor, and virtually all the volts are dropped across the resistor! Without a current limiting resistor the bands flatten out, enormous currents flow, and the device is destroyed. Let's summarize what we've covered here so far.

Forward bias:

1 Junction potential reduced.
2 Enhanced hole diffusion from p-side to n-side compared with the equilibrium case.
3 Enhanced electron diffusion from n-side to p-side compared with the equilibrium case.
4 Drift current flow is similar to the equilibrium case.
5 Overall, a large diffusion current is able to flow.
6 Mnemonic. Connect positive terminal to p-side for forward bias.

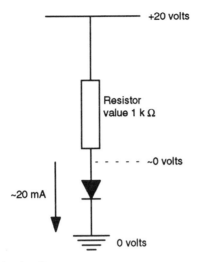

+20 volts

Resistor
value 1 k Ω

~0 volts

~20 mA

0 volts

Figure 6.15   Simple diode circuit.

Reverse bias:

1 Junction potential increased.
2 Reduced hole diffusion from p-side to n-side compared with the equilibrium case.
3 Reduced electron diffusion from n-side to p-side compared with the equilibrium case.
4 Drift current flow is similar to the equilibrium case.
5 Overall a very small reverse saturation current flows.
6 Mnemonic. Connect positive terminal to n-side for reverse bias.

6.7   Qualitative explanation of forward bias

So far we've come quite a long way with the p–n junction with a minimum of equations and we've got quite a good idea of what's going on electrically. However, our picture of the diode up to this point has been quite simple, and, like most topics when you start to think a little bit about what's going on, things suddenly start getting quite difficult. If you can't see what's looming, let's start off by thinking about the p–n junction under forward bias conditions. The first bit is easy: the forward bias lowers the junction potential and large diffusion currents can flow. But now we start to think about what is physically going on and things suddenly get complicated. By forward biasing the diode I have *injected* a large number of holes from the p-side, across the depletion region, and they now find themselves on the n-side of the junction. Similarly, with forward bias I've injected a large number of electrons from the n-side across the depletion region, and they now find themselves on the p-side of the junction. The situation is shown in Figure 6.16. We're still not quite stuck yet because we know how to handle this situation from our work on carrier diffusion (see Section 4.12). We know that this excess of injected minority carriers will recombine with the majority carriers and that the minority carrier excess will decay away exponentially with distance into the majority carrier side of the junction as shown in Figure 6.16. By the time we've got a few minority carrier diffusion lengths away from the depletion-layer edge the excess minority carrier concentration will have dropped to the equilibrium minority carrier concentration in the neutral material. So how does any current flow? If *all* the injected minority carriers recombine with the vast number of majority carriers within a few diffusion lengths, how *does any* current flow? We need to go into an extra depth of complexity. We now need to consider the role of the *ohmic contacts* at the ends of the device. Yes, the injected minority carriers *do* recombine with the majority carriers *but* if we are to maintain *charge neutrality* the majority carrier that has gone missing (by recombining with an injected minority carrier) needs to be replaced, and that's exactly what happens. *The 'lost' (recombined) majority carriers are replaced by (majority) carriers coming in from the ohmic contacts, thus maintaining charge neutrality.* The sum of the hole and electron currents flowing through the ohmic contacts makes up the total current flowing through the external circuit! Referring to Figure 6.16 again and looking at just the p-side for a moment, most of the current is carried by *electrons* near the depletion-layer edge, but as we go into the neutral material, more and more current is carried by the holes until we get far away from the depletion region where at the metal contact the current is purely a hole current. The converse situation applies for the n-side of the junction.

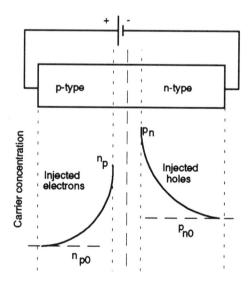

Figure 6.16 p–n junction in forward bias.

I did say that the p–n junction was a complicated device! With bipolar conduction you have four balls in the air at once! Qualitatively we now have some idea of what's going on. It's time to formulate a quantitative explanation.

6.8 The ideal diode equation

We need to introduce some new nomenclature. First refresh your memory by looking at the end of Section 3.1 to remind yourselves of minority/majority carrier concentrations in n/p-type material. The additional piece of nomenclature we need to introduce is a zero subscript which denotes *equilibrium* concentrations. We thus obtain the following four quantities:

(a) n_{n0} equilibrium electron concentration in n-type material,
(b) n_{p0} equilibrium electron concentration in p-type material,
(c) p_{p0} equilibrium hole concentration in p-type material,
(d) p_{n0} equilibrium hole concentration in n-type material.

The equation that we encountered way back in Chapter 3 namely,

$$n \cdot p = n_i^2 \tag{3.2}$$

can therefore be more formally written using the new nomenclature as

$$\left.\begin{aligned} n_{n0} \cdot p_{n0} = n_i^2 \quad \text{n-type material} \\[2mm] p_{p0} \cdot n_{p0} = n_i^2 \quad \text{p-type material} \end{aligned}\right\} \tag{6.37}$$

We could therefore write the equation for the built-in voltage more precisely as

$$V_{bi} = \frac{kT}{q} \ln\frac{n_{n0}p_{p0}}{n_i^2} \tag{6.38}$$

which, using equation (6.37), gives

$$V_{bi} = \frac{kT}{q} \ln\frac{n_{n0}}{n_{p0}} = \frac{kT}{q} \ln \cdot \frac{p_{p0}}{p_{n0}} \tag{6.39}$$

which on rearranging gives

$$n_{n0} = n_{p0} \exp\left(\frac{qV_{bi}}{kT}\right) \tag{6.40}$$

and,

$$p_{p0} = p_{n0} \exp\left(\frac{qV_{bi}}{kT}\right) \tag{6.41}$$

We now need to start setting up the equations for the p–n junction diode under forward and reverse bias conditions. If we apply forward bias the potential difference reduces to $V_{bi} - V_F$, where V_F is the forward bias voltage. For reverse bias the potential difference increases to $V_{bi} + V_R$, where V_R is the reverse bias voltage. With these applied biases the carrier densities change from equilibrium carrier densities to non-equilibrium carrier densities. If we consider forward bias for the moment, then the non-equilibrium forward bias condition for the electrons is simply extracted from the equilibrium condition (equation (6.40)) as

$$n_n = n_p \exp\left(q\frac{[V_{bi} - V_F]}{kT}\right) \text{ NON-EQUILIBRIUM} \tag{6.42}$$

We now make a simplifying assumption and consider *low-level injection* only. For low-level injection the injected *minority* carrier density is very much less than the *majority carrier* density. In other words the injected minority carriers do not upset the majority carrier equilibrium densities giving us the low-level injection conditions

$$n_n \approx n_{n0} \text{ LOW-LEVEL INJECTION} \tag{6.43}$$

and,

$$p_p \approx p_{p0} \text{ LOW-LEVEL INJECTION} \tag{6.44}$$

If we put the low-level injection result of equation (6.43) into equation (6.42) we obtain

$$n_{n0} = n_p \exp\left(\frac{q[V_{bi} - V_F]}{kT}\right) \tag{6.45}$$

Now substitute for n_{n0} using equation (6.40) to obtain

$$n_{p0} \exp\left(\frac{qV_{bi}}{kT}\right) = n_p \exp\left(q\frac{[V_{bi} - V_F]}{kT}\right) \tag{6.46}$$

A slight rearrangement gives

$$n_p = n_{p0} \exp\left(\frac{qV_F}{kT}\right) \tag{6.47}$$

Which on subtracting n_{p0} from both sides finally gives

$$n_p - n_{p0} = n_{p0}\left[\exp\left(\frac{qV}{kT}\right) - 1\right] \tag{6.48}$$

A similar analysis for the holes yields

$$p_n - p_{n0} = p_{n0}\left[\exp\left(\frac{qV}{kT}\right) - 1\right] \tag{6.49}$$

The subscript F has been removed from equations (6.48) and (6.49) since these equations hold good for reverse bias as well, provided we replace V with $-V$.

These equations help us further with our picture of the p–n junction diode. These equations give us the *excess* concentration of minority electrons and holes (over the equilibrium concentration) at the edges of the depletion region, so this gives us two extra points on our diagram (A and B). More importantly, you will remember that it is this *excess* minority carrier concentration that is the driving force for the *diffusion* process, so that very soon we will be able to couple these results with the diffusion equations.

Let's put all these results together in a diagram, as illustrated in Figure 6.17. This diagram is for the diode in forward bias so we have the injection of minority

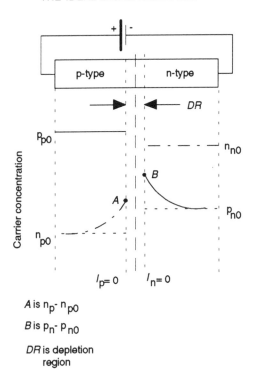

Figure 6.17   p–n junction in forward bias.

carriers across the depletion region. Taking the *neutral p-region* first we have a thermal equilibrium concentration of the majority carrier holes, p_{p0}, and a thermal equilibrium concentration of the minority carrier electrons n_{p0}. Under forward bias the majority carrier holes are injected across the depletion region to appear as *minority* carriers at the edge of the depletion region on the n-side. The concentration of these minority carrier holes at point B is what we've just worked out in equation (6.49). These minority carrier holes will diffuse in the *field-free n-region* (there is no electric field *but* there is a carrier concentration gradient) towards the ohmic contact, which is many diffusion lengths away to the right. Since the ohmic contact is a long way away, the minority hole concentration decays exponentially with distance into the n-region until it reaches the minority carrier thermal equilibrium concentration p_{n0}. For simplicity, I have taken the origin for the n-side to occur at the depletion-layer edge.

If we now look at the *neutral n-side*, the thermal equilibrium concentration of majority carrier electrons is n_{n0}, and for the minority carrier holes it is p_{n0}. The majority carrier electrons are injected across the depletion region to appear as *minority* carriers at the edge of the depletion region on the p-side. This minority carrier electron concentration decays from the value we calculated from equation (6.48) at point A, to the thermal equilibrium value n_{p0}, as we travel towards the p-side ohmic contact way to the left of the diagram. Again for convenience I have chosen the origin for the p-side of the diode to be at the depletion-layer edge as well. Note at this point that I have *not* drawn any lines indicating what might be going on *in* the depletion region. We've got no mathematical information on what's occurring in here, although we can reasonably assume current continuity (no sources or sinks of current) across the depletion region. From our knowledge of the diffusion equations we can now pull all this information together to get at the ideal diode equation.

We now need to start using the diffusion equations of Chapter 4. Although we have the excess minority carrier densities at the edge of the depletion region, we haven't written down the full equation which includes the carrier decay with distance into the semiconductor. If you return to Section 4.12 and look at equations (4.66) and (4.67) you'll see that we can easily take this decay into account by simply multiplying by an exponential function. So if we take our excess minority carrier concentration equations (6.48) and (6.49) and couple these with equations (4.66) and (4.67) we'll get

$$\delta n(l_p) = \delta n(0) \exp\left(-\frac{l_p}{L_n}\right) \tag{6.50}$$

for the electrons, and

$$\delta p(l_n) = \delta p(0) \exp\left(-\frac{l_n}{L_p}\right) \tag{6.51}$$

for the holes. In case it's not completely clear,

$n_p - n_{p0} = \delta n$
$p_n - p_{n0} = \delta p$
l_p is the distance into the p-region from the depletion-region edge,
l_n is the distance into the n-region from the depletion-region edge,
$\delta n(0) = n_{p0}(\exp[qV/kT] - 1)$,
$\delta n(0) = p_{n0}(\exp[qV/kT] - 1)$.

As we now have the minority carrier concentrations as a function of distance all we need to do is use equations (4.44) and (4.45) to convert equations (6.50) and (6.51) into electron and hole *current densities*. The total current density flowing through the diode will be the sum of these two quantities. Let's work through the electron current density. We start with equation (4.44) for the diffusion current density:

$$J_n = qD_n\frac{dn}{dx} \tag{4.44}$$

We then apply equation (4.44) to the electron excess minority carrier density equation (6.50), which may be written in full as

$$\delta n(l_p) = n_{p0}\left(\exp\left[\frac{qV}{kT}\right] - 1\right)\exp -\frac{l_p}{L_n} \qquad (6.52)$$

If we perform the differentiation and calculate the current density at $x_p = 0$ we obtain

$$J_n(l_p = 0) = -\frac{qD_n n_{p0}}{L_n}\left(\exp\left[\frac{qV}{kT}\right] - 1\right) \qquad (6.53)$$

The minus sign in equation (6.53) means that the electron current is in the opposite direction to increasing l_p; in other words it is in the positive x-direction. Exactly the same analysis gives the hole current density at $l_n = 0$:

$$J_p(l_n = 0) = -\frac{qD_p p_{n0}}{L_p}\left(\exp\left[\frac{qV}{kT}\right] - 1\right) \qquad (6.54)$$

This current is in the positive x-direction as well. Adding equations (6.53) and (6.54) together gives us the total current density due to electrons *and* holes flowing through the diode as

$$J_{TOTAL} = q\left(\frac{D_p p_{n0}}{L_p} + \frac{D_n n_{p0}}{L_n}\right)\left(\exp\left[\frac{qV}{kT}\right] - 1\right) \qquad (6.55)$$

which may be written in the more familiar form

$$J_{TOTAL} = J_0\left(\exp\left[\frac{qV}{kT}\right] - 1\right) \qquad (6.56)$$

Finally, if we simply multiply through by the diode area to convert current densities into currents, we get the familiar form of the ideal diode equation:

$$I = I_0\left(\exp\left[\frac{qV}{kT}\right] - 1\right) \text{ IDEAL DIODE EQUATION} \qquad (6.57)$$

This equation is good for reverse as well as forward currents. All we need to do is change the sign of V! If we do this and make sure that qV is greater than a few kT then the exponential factor tends to zero and we get

$$I = -I_0 \text{ REVERSE SATURATION CURRENT} \qquad (6.58)$$

as we would expect from the diode $I–V$ characteristic. In the derivation, one point I have 'glossed over' is that I've assumed that the current density of the electrons is the same at both edges of the depletion region, and I have assumed the same thing for the holes. What this means physically can be seen in Figure 6.18, where there are two horizontal lines (electron and hole current density) *through* the diode depletion region. The assumption, as you see, implies current *continuity* *through* the depletion region with no generation or recombination in the depletion region itself. In other words, all injected carriers get across! Note also the *constant* current density $J_n + J_p$ flowing through the diode. Although the minority carrier current density may be exponentially *decreasing* with distance into the neutral regions, the majority carrier current density is exponentially *increasing*. Summing these two current density components gives the constant current density as indicated in Figure 6.18.

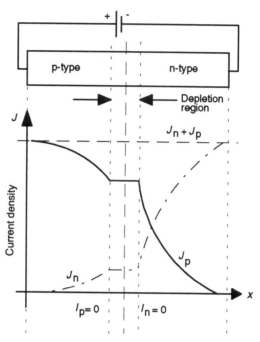

Figure 6.18 Forward bias current densities.

Because we've spent such a long time looking at forward bias I will only have a quick look at the reverse bias condition. With the background you've gained from the above work you should be able to make a more quantitative analysis of reverse bias if you want to. Figure 6.19 shows the carrier density and current density through the diode under reverse bias conditions. Notice the wider depletion-layer width under reverse bias! Although the application of a reverse bias prevents the large forward bias diffusion currents from flowing, it doesn't prevent all current flow. Reverse bias allows minority carrier holes to be extracted from the n-region and minority carrier electrons from the p-region. In fact, the electric field near the junction is large enough to extract all the minority carriers near the junction. This flow of minority carriers across the junction constitutes I_0, the reverse saturation current. On average all the minority carriers (which are being thermally generated) within distances L_p and L_n from the edge of the depletion region are extracted by the field across the junction. This situation is shown in the carrier density profile in Figure 6.19. The reverse saturation current density will increase exponentially with *temperature* but it is *independent* of the *reverse bias* provided we don't start getting generation centres in the enlarged depletion region and provided we don't start getting *junction breakdown*.

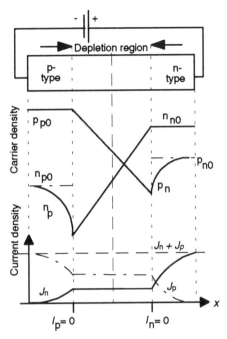

Figure 6.19   p–n junction in reverse bias.

Calculation

Calculate the forward voltage drop V_F of an ideal p–n junction diode at room-temperature for a forward current of 15 mA if the reverse saturation current is

(a) 5 μA,
(b) 8 pA.

If these two reverse saturation currents are measured for two different diodes having exactly the same physical dimensions, which is the silicon diode and which is the germanium diode?

To find the forward voltage drop we simply rearrange the ideal diode equation (6.57):

$$I = I_0\left(\exp\left(\frac{qV_F}{kT}\right) - 1\right)$$

Rearrange to get

$$\ln\left(\frac{I}{I_0} + 1\right) = \frac{qV_F}{kT}$$

Making V_F the subject of the equation finally gives

$$V_F = \frac{kT}{q} \ln\left(\frac{I}{I_0} + 1\right)$$

If we assume the usual value of 25 mV for kT/q at room-temperature, then substituting values into the equation for V_F gives

(a) 0.2 volt,
(b) 0.53 volt.

The reverse leakage current is due to the thermally generated minority carriers near the edge of the depletion region being swept across the region by the in-built junction field. From our Chapter 3 work we know that *more* minority carriers will be thermally generated by the semiconductor with the *smaller* energy gap. The higher reverse leakage current must therefore be the germanium diode, and the smaller leakage current corresponds to the silicon diode. Hence,

(a) germanium diode,
(b) silicon diode.

Notice that an equivalent way of looking at this is to say that the larger band gap material has the larger forward voltage drop across it, for the same forward current, all other things being equal. This unavoidable forward voltage drop with forward current is of concern to *circuit designers*, who must take this into account in their designs.

6.9 Reverse breakdown

Figure 6.20 shows what happens to the p–n junction *I–V* characteristic at large reverse biases. The diode breaks down and large currents start to flow once we've hit the diode's reverse breakdown voltage V_B. If we have a current limiting resistor in series with the diode, we can control the current to a safe value, and prevent the destruction of the device. If there is no current limiting resistor, a large uncontrolled current will flow on reaching the breakdown voltage, and the diode will be destroyed. Some diodes are made to have specific breakdown voltages to be used as *voltage reference diodes* in circuits where there is a controlled current flow. The breakdown mechanism depends on the type of diode. For highly doped p–n junctions the breakdown mechanism is *Zener breakdown*, which usually occurs for voltages of about 5 V or less. For less highly doped diodes the breakdown mechanism is *avalanche breakdown*, and this is used for higher voltage reference diodes. In both cases, however, these diodes are sold as *Zener* (voltage reference) *diodes*.

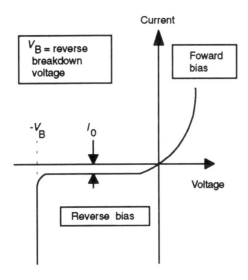

Figure 6.20   Reverse breakdown.

Calculation

Sketch the electric field in a p–n junction in thermal equilibrium with a doping level of $10^{15}\,\mathrm{cm}^{-3}$ on each side of the junction. Show on the same sketch how the electric field changes with increasing reverse bias. Calculate the maximum electric field at the junction if the diode is silicon and 150 V reverse bias is applied. Neglect the built-in potential for this calculation. Calculate the breakdown voltage of the same diode if breakdown occurs when an electric field of $10^5\,\mathrm{V/cm}$ is reached.

Figure 6.21 shows how the electric field increases at the junction with increasing reverse bias. Remember what is happening from a physics point of view. As the reverse bias is increased the depletion-region width increases and more charge is uncovered. More uncovered charge means that a greater electric field is produced (from Poisson's equation).

For the maximum electric field, go back to equation (6.20):

$$\mathbf{E}_m = -\frac{qN_D x_n}{\varepsilon_{Si}} = -\frac{qN_A x_p}{\varepsilon_{Si}} \tag{6.20}$$

and note that since the doping on either side of the junction is the same, the width of x_n will be the *same* as the width of x_p. The voltage across the junction is simply minus the integral of the electric field with respect to distance, or, more simply, the area under the triangle. Hence,

$$\text{VOLTAGE} = \mathbf{E}_m \frac{|x_n| + |x_p|}{2} = \mathbf{E}_m |x_n|$$

Neglecting the built-in voltage means that this voltage across the junction is 150 V, so

$$V = 150\,\text{V} = \mathbf{E}_m \frac{\varepsilon_{Si}}{qN_D} \mathbf{E}_m$$

where I have substituted for x_n. It just remains to solve the equation for $\mathbf{E}_m$:

$$\mathbf{E}_m = \sqrt{\frac{150qN_D}{\varepsilon_{Si}}}$$

Substituting values finally gives

$$\mathbf{E}_m = 1.5 \times 10^7\,\text{V}\,\text{m}^{-1}$$

If breakdown occurs when the electric field is only 10^5 V/cm, then rearrange the equation to get

$$V = \mathbf{E}_m^2 \frac{\varepsilon_{Si}}{qN_D}$$

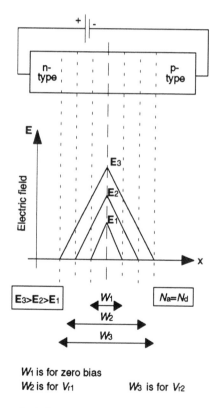

W₁ is for zero bias
W₂ is for V_{r1}   W₃ is for V_{r2}

Figure 6.21   Diagram for calculation.

and substitute values to obtain

$$V = 64.7 \text{ volts (breakdown voltage)}$$

From the equations it should be clear that a lower doping implies a higher breakdown voltage. Physically this fits in OK since it means a given voltage will produce a bigger depletion-region width in a lower doped junction, hence the volts will be dropped over a greater length, hence the electric field is smaller. It should also be clear that a one-sided abrupt junction will break down at half the volts of a symmetrical junction which has the same doping as the low-side of the abrupt junction. Hence for higher breakdown voltages go for symmetrical junctions rather than one-sided junctions.

Avalanche breakdown occurs because the electric field across the junction has become so large, due to the applied reverse bias, that an electron can gain enough energy to break a bond when it collides with a lattice atom. Physically, this means that the electron can ionize an atom (promote an electron from the valence to the conduction band) and generate an electron–hole pair. The electron–hole pair is now itself accelerated, generating more electron–hole pairs in the process. This mechanism is called avalanching and is also known as *impact ionization*.

Zener breakdown, in contrast to avalanche breakdown, requires a highly doped p–n junction. In such highly doped junctions the depletion width is already very small so even without any bias a very high electric field exists at the junction. This being the case, only very small reverse biases are necessary to increase the electric field to the point where electrons are *directly removed* from their valence bonds. This mechanism is Zener breakdown and typically occurs for low voltages of less than 5 volts.

6.10 Depletion capacitance

The final property of a p–n junction that I wish to cover in this chapter is that of junction (depletion) capacitance.

A capacitor is a passive circuit element which can store charge according to

$$Q = CV \qquad\qquad (6.59)$$

where

Q is the stored charge in Coulombs,
C is the capacitance in Farads,
V is the voltage across the capacitor.

When we apply reverse bias to the p–n junction we uncover more charge in the depletion region (compared to the thermal equilibrium condition). Similarly, when we apply a forward bias to a diode the depletion region is reduced, and so is the amount of uncovered charge. Since we clearly have a device which has an amount of charge in its depletion region which depends on the voltage across the device, we have something that must look like a capacitor. In fact, since the capacitance of a diode depends on the width of the depletion region, and the depletion-region width depends on the applied bias, we have a *variable capacitance* element whose capacitance depends on bias. This variable capacitance property of the p–n junction is used in some devices that are fabricated specifically for this purpose. These variable capacitance or *varactor* diodes are often used in frequency tuning circuits. It turns out that the capacitance per unit area of diode is given by

$$C_{DEP} = \frac{\varepsilon_{Si}}{W} \, F \, cm^{-2} \qquad (6.60)$$

The form of equation (6.60) is exactly the same as for a parallel plate capacitor! I now want to look at the variable capacitance properties of the one-sided abrupt junction since it has a very interesting application.

The depletion-layer width for a one-sided abrupt junction in thermal equilibrium is given by equation (6.30). By now we know how to include the effect of reverse bias and so we get

$$W = \left(\frac{2\varepsilon_{Si}[V_{bi} + V_R]}{qN_D} \right)^{\frac{1}{2}} \qquad (6.61)$$

From equation (6.60) we can immediately write down the capacitance per unit area:

$$C_{DEP} = \left(\frac{q\varepsilon_{Si}N_D}{2[V_{bi} + V_R]} \right)^{\frac{1}{2}} \qquad (6.62)$$

Finally, equation (6.62) is rearranged to give

$$\frac{1}{C^2} = \frac{2(V_{bi} + V_R)}{q\varepsilon_{Si}N_D} \qquad (6.63)$$

In the above I'm choosing the n-side to be low-doped; we could have chosen the p-side had we wished. However, equation (6.63) provides a very interesting result. If we were to take measurements of a one-sided abrupt junction in a circuit that measures both diode capacitance and applied voltage (which is easily done), then we could make a graph of equation (6.63). If we plotted $1/C^2$ against V_R we would get a slope from which we could extract N_D, and an intercept giving us V_{bi}!! This is very useful information! If we have a piece of semiconductor material, and we don't know its doping concentration, we could perform the above capacitance–voltage (C–V) experiment to find this information out. Figure 6.22 shows C–V data for a p–n junction diode. The data has been linearized by plotting $1/C2$ against V. Using Figure 6.22, and applying equation (6.63), we can obtain the doping density on the low-doped side of the diode (from the slope of the line), and we can obtain the built-in voltage from the intercept of the line with the x-axis.

Calculation

Find the built-in voltage and the doping concentration on the low-doped side of the $p^{++}n$ diode from the $1/C^2$ versus V plot shown in Figure 6.22. The diode area is $4.0 \times 10^{-5}\,\mathrm{m}^2$.

From equation (6.6) the intercept on the voltage axis gives the voltage at which $V_R = -V_{bi}$ (since $1/C^2$ is zero on the voltage axis). Since the intercept occurs at $-0.5\,\mathrm{V}$, the built-in voltage $V_{bi} = 0.5\,\mathrm{V}$.

Don't forget that the capacitance C in equation (6.63) is the capacitance *per unit area*. Converting equation (6.63) to an equation containing the capacitance we get

$$\frac{1}{C_T^2} = \frac{2(V_{bi} + V_R)}{q\varepsilon_{Si}N_d A^2}$$

where

C_T is the capacitance (Farads),
A is the device area.

Hence the slope of the graph in Figure 6.22 is

$$SLOPE = \frac{2}{q\varepsilon_{Si}N_d A^2}$$

from which we get the doping density on the n-side of the diode, N_d, given by

$$N_d = \frac{2}{q\varepsilon_{Si}\, SLOPE\, A^2}$$

and substituting constants together with the slope of the graph gives

$$N_d = 4.35 \times 10^{20}\,\mathrm{m}^{-3}$$

or,

$$N_d = 4.35 \times 10^{14}\,\mathrm{cm}^{-3}$$

Why should we have a piece of material and we don't know the doping? We may have just grown it ourselves in a chemical vapour deposition or molecular beam epitaxy apparatus. We may have done more than just grow a piece of uniformly doped silicon though; we may have changed the doping concentration as we grew the layer. Could we measure this with our C–V apparatus? Yes we can, but things get slightly more involved. What we now want to measure is doping concentration with depth, the doping *profile* in the semiconductor. To do this we apply a small reverse bias and then on top of this we apply a small a.c. signal. The small reverse bias pushes the depletion-layer edge a little way into the semiconductor and the small a.c. signal 'jiggles' the edge of the depletion region, which allows us to apply equation (6.63) and *measure the capacitance* and hence the *doping concentration at the edge of the depletion region*. We can now increase

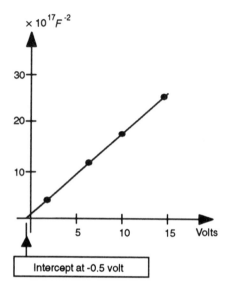

Figure 6.22   Linearized diode $C–V$ characteristic.

the d.c. reverse bias, push the depletion-layer edge in a little further, and measure the doping concentration again. In this way we can build up the doping profile of the semiconductor sample! But wait a minute. This can't keep going on. We've completed a whole section on breakdown voltage and we know that eventually the diode will break down under the reverse bias. This breakdown may well occur before you have profiled into the semiconductor far enough! This is true, but there is a really good way out of this problem: *electrochemical C–V profiling*.

If we could find a chemical solution that made a good rectifying contact to our unknown silicon sample, then we could put our reverse bias on this chemical contact and push the depletion region into the semiconductor, and then make our capacitance measurement as in Figure 6.23. An added bonus would be that we wouldn't have to use special fabrication techniques to make a diode. We would just put our piece of silicon in a cell with this liquid and start measuring! Fortunately, such chemical solutions do exist. For silicon measurements ammonium bifluoride solution can be used. But the good news doesn't stop there. Remember, we will still have the diode breakdown problem. What can be done about this? Amazingly, by applying a slightly different bias to the solution we can actually *etch away* some silicon (in a very controlled way) before making our capacitance measurements! Proceeding in this way we can etch some silicon, make our capacitance (doping concentration) measurement, and then etch away some more silicon. We can get a doping profile through the sample, etching the silicon away as we go, without diode breakdown problems. We do, however, destroy the small piece of silicon on which we are making the measurement. It is because the etching is an electrochemical process that we can accurately measure how much silicon is removed. If we know the equations for the electrochemical reaction, and we accurately measure the etching current, then provided we know the contact area we're etching very accurately, we can calculate the depth of silicon we're removing! This is a useful characterization technique based on our knowledge of the p–n junction diode!

Calculation

Calculate the junction (depletion) capacitance of a silicon p^+n junction of area $1.14 \times 10^{-3}\,cm^2$ if there is a reverse bias of 10 volt and the doping is $10^{16}\,cm^{-3}$ on the n-side.

The relevant equation is given by (6.62):

$$C_{DEP} = \left(\frac{q\varepsilon_{Si}N_D}{2[V_{bi} + V_R]}\right)^{\frac{1}{2}}$$

which gives the capacitance per unit area.

If we choose a high doping for the p^+ region of, say, $10^{20}\,cm^{-3}$ or more, then using equation (6.11) for V_{bi} you should get a voltage approaching unity. Check this for yourself using several high doping values. I shall use a value of 1 volt for V_{bi} and accept the small error. Now just substitute values into the above equation to get

$$C_{DEP} = \left(\frac{1.6 \times 10^{-19} \times 11.9 \times 8.854 \times 10^{-14} \times 10^{16}}{2(1 + 10)}\right)^{\frac{1}{2}} F\,cm^{-2}$$

$$= 8.754 \times 10^{-9}\,F\,cm^{-2}$$

Finally, multiplying by the device area gives a junction capacitance of

$$C_j = 8.754 \times 10^{-9} \times 1.14 \times 10^{-3} = 10\,pF$$

Another long chapter! The *really* bad news is that we *still* don't know all about the p–n junction. We haven't looked at the following:

(a) the stored charge (the injected minority carriers) and how it changes with time,
(b) the derivation of the depletion capacitance,
(c) The physics of an ohmic contact,
(d) generation and recombination *in* the depletion region,
(e) graded junctions,

to name just a few. However, we *have* covered enough material to press on further.

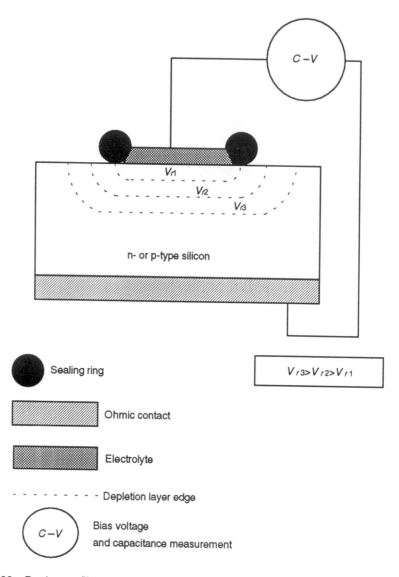

Figure 6.23   Doping profile apparatus.

CHAPTER SEVEN

LED, photodetectors and solar-cell

The next decade or so will prove a very interesting time, when photons as information carriers will be far more important than electrons or holes. By this statement I do not mean the photon information carriers we are already well-acquainted with, radio waves, since these low-energy electromagnetic waves are already predominant. I mean the transfer of information within electronic circuits themselves! Photons have many properties that make them far more suitable as information carriers, within large-scale integrated circuits, than either electrons or holes.

The different parts of an integrated circuit (IC) are connected together by metal (including metal alloy and metal multilayers), silicide, or highly doped polysilicon conducting tracks. These tracks (or interconnects) also take voltages and signals to the chip from external sources, or, conversely, they carry signals from the chip to the outside world. For connection to the outside world, the interconnects terminate on on-chip *bonding pads*, and these bonding pads can be accessed by the circuit designer via the metal legs that stick out of the side of chip packages. High current densities of either electrons or holes in integrated circuit interconnections can break the current carrying tracks. It is clear that simple ohmic heating from high current densities could break the tracks, but even before this occurs, tracks can be broken by *electromigration*. The electrons and holes can carry considerable momentum at high velocities which can be transferred to the track's lattice atoms during collisions. This momentum transfer is capable of physically moving track atoms which can eventually cause a break in the track (it can also cause a short circuit with a close neighbouring track). High densities of *photons* moving down their equivalent tracks (waveguides) don't cause such problems. Recall that photons carry very little momentum.

As a second example of the superiority of photons for this information carrying role, consider high-frequency signals moving down tracks and waveguides. High-frequency electron or hole currents flowing down tracks radiate energy proportional to the fourth power of the frequency! Clearly, we are going to have a lot of trouble trying to send lots of high-frequency signals down lots of closely packed tracks. All that radiated energy can be picked up by close tracks causing *crosstalk*, information from one track getting into a neighbouring track. Photons, on the other hand, do not give us such frequency-related crosstalk problems. The photon is already oscillating at an extremely high frequency, and waveguides can be made that produce much less optical crosstalk than their electrical counterparts. There are many more reasons why we might choose photons, rather than electrons or holes, but those two examples will serve us for now.

Having seen the potential use of photon information carriers we now need a photon source to produce the information, and a photon detector to detect the information. Solid-state photon sources include the light emitting diode (LED) and the semiconductor LASER. We shall look at the fundamentals of the LED in this chapter. There are many different solid-state photon detectors available, but as this chapter follows work on the p–n junction, we shall concentrate on junction detectors. Having looked at p–n junction photodetectors, an obvious thought should come to mind. The LED provides us with a device that will produce *photons* if we supply some *current*. Does a device exist that produces *current* if we supply it with *photons*? Of course it does, it's the photocell, or solar-cell if the photon source is sunlight. This set of devices forms a logically consistent group for study in this chapter.

7.1 The light emitting diode

The light emitting diode structure is shown in Figure 7.1. You can see that the LED is little more than a diode fabricated in a direct band-gap material. We know from the work in Chapter 2 that in order to achieve a reasonable efficiency for photon emission, the semiconductor must have a direct band gap. However, this does not prevent us from using indirect band-gap materials (including modified indirect gap materials) for LEDs operating at a lower efficiency. We shall start off by looking at the mechanism for photon emission in these devices, and then take a closer look at the materials used in their construction.

The term used to describe the mechanism behind photon emission in LEDs is *injection electroluminescence*. The *luminescence* part tells us that we're producing photons. The *electro* part says that the photons are being produced by an electric current. Finally, *injection* tells us that the photon production is by the injection of current carriers. The LED is therefore a device that produces photons via a current injection mechanism. We know that a photon is produced when an electron recombines with a hole across the energy gap of a direct band-gap semiconductor. We also know that the energy of the photon produced in this way is given by the energy gap of the semiconductor. All we need now is a method of putting a lot of electrons in a region where there are a lot of holes (or vice-versa), so that many electron–hole recombination events can occur, and lots of photons can be produced. This is where the diode structure comes in. We know from the last chapter that *forward biasing* a p–n junction will inject lots of electrons from the n-side, across the depletion region and into the p-side *where they will combine*

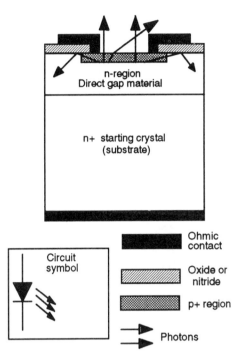

Figure 7.1   Basic LED structure.

with the high density of majority carriers. Similarly, holes will be injected into the n-side of the junction from the p-side, and recombination will again result. Provided the junction is formed in a direct band-gap semiconductor this recombination of carriers will result in efficient photon production. Figure 7.2 shows us what is going on around the junction region. Notice that we will get photon emission wherever we have injected minority carriers recombining with the majority carriers. This means that *if* the electron diffusion length is greater than the hole diffusion length, then the photon emitting region will be bigger on the p-side of the junction than on the n-side, as shown in Figure 7.2 (I've made the depletion region infinitely thin in this diagram). What this means when it comes to constructing a real LED in such material is that it may be best to consider a $n^{++}p$ structure. Why? Because the n^{++} contact will inject plenty of electrons into the p-region for recombination, and the longer electron diffusion length means a larger emitting volume on the p-side of the junction. This is a rather simplistic approach to LED construction and in practice you would also have to consider the photon production efficiencies of the n-type material, and the p-type material, which may differ substantially. However, it is usual to find the photon emitting volume occurs mostly on one side of the junction region, and this applies to LASER devices as well as LEDs (for the same reasons).

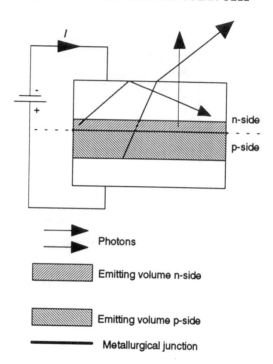

n-side

p-side

→▶ Photons

Emitting volume n-side

Emitting volume p-side

——— Metallurgical junction

Figure 7.2 LED section near the junction.

7.2 Materials for LEDs

The semiconductor band-gap energy defines the energy of the emitted photons in an LED. If we are to fabricate LEDs that can emit photons from the infrared to the ultraviolet parts of the electromagnetic spectrum, then we must consider several different materials systems. No single system can span this energy band at present, although the 3–5 nitrides come close. We have further constraints, which are covered in more detail in Chapter 10, but which I will mention briefly here. If we are to make a p–n junction, then our semiconductor must be able to be doped both p- and n-type. Unfortunately, many of the potentially useful 2–6 group of direct band-gap semiconductors (ZnSe, ZnTe, etc.) come naturally doped *either* p-type, *or* n-type, but they do not like to be type-converted by overdoping. The material reasons behind this are complicated and are not entirely well-known, but very recently blue LEDs and LASERS *have* been fabricated in the 2–6 materials with much difficulty, after a couple of decades of diligent research! The same problem is encountered in the 3–5 nitrides and their alloys InN, GaN, AlN, InGaN, AlGaN, and InAlGaN. The amazing thing about the 3–5 nitride *alloy* systems is that they appear to be direct gap throughout the full composition range! Since InN has a direct gap of about 2 eV, and GaN has a direct gap of about 3.3 eV, the InGaN system can cover most of the visible spectrum!

Calculation

Over what range of photon wavelengths could the InGaN alloy system emit photons? What colour do these wavelengths correspond to?

Before carrying out the calculation I need to clear up one point. When we consider the InGaN alloy we will consider one composition at a time only, say 10% Ga, 40% In, and 50% N. So this material will emit *one wavelength only* corresponding to this particular composition. An InGaN LED would not emit white light (the whole of the visible spectrum at once) because the InGaN in which we form our junction would have a specific composition. It would be possible, in theory at least, to form a complicated multilayer device emitting lots of different wavelengths which would look like a white light source at a distance. However, when we looked closely at such a device we would see that each layer, with its specific composition, would be emitting a specific wavelength. We could go even further into the realms of science fiction and imagine an LED fabricated in *graded material* where on either side of the junction region the material changes slowly from InN to GaN via InGaN alloys. This would mean the injected minority carriers need to get through the whole of this alloy region if efficient photon production at all visible wavelengths was to occur. With this device I'm also ignoring the horrific materials and growth problems! Finally, although this materials system appears ideal, it's unfortunately got the same problem as a lot of the 2–6 compound semiconductors: it comes n-type and doesn't like to be doped p-type. Having said that, very recently a Japanese group managed to type-convert some GaN, and they made an LED in the material. Let's return to the original question and see what 'colour' photons such an LED would produce.

Starting with the short wavelength end first (i.e. the biggest band gap) we'll consider the highly gallium rich alloy, in other words GaN. With a 3.3 eV band gap the photons will have a wavelength of 376 nm (using equation (2.2)), which puts them just in the ultraviolet part of the electromagnetic spectrum. The Japanese LED would therefore emit ultraviolet photons, which it did. Now move on to the indium-rich alloy InN with a band gap of 2 eV. This energy corresponds to a photon wavelength of 620 nm, which is in the visible part of the electromagnetic spectrum and is associated with *orange* light.

The InGaN alloy system could therefore cover most of the visible spectrum, as an LED (and LASER) material, were it not for the doping difficulties mentioned above. Contained in the above calculation is a result I find truly amazing. The calculation shows that the *whole* of the visible spectrum is covered by just 1 eV of energy. The most energetic gamma rays (cosmic rays) have photon energies of about 10^9 eV. The weakest energy photons, corresponding to very long wavelength radio waves, have photon energies of about 10^{-11} eV or less. The electromagnetic spectrum itself therefore covers something in excess of 10^{20} eV, and yet our eye can only detect photon energies between 2 and 3 eV!!

I've been diverted and looked at exciting direct-gap materials that aren't commercially available at present. It's time to return to reality and see what materials systems are used in the LEDs you are able to buy today.

Starting with the long wavelength (infrared) emitters first, it is clear that the ubiquitous AlGaAs/GaAs system will make an appearance. Why? Because we know that we can dope the system both p- and n-type (and therefore make diodes), and from Chapter 2 we know that AlGaAs is direct gap for *some* aluminium concentrations (and therefore a good emitter of photons). What wavelength range can we cover using this system? From a calculation in Chapter 2 we know that the band gap of GaAs is about 1.43 eV at room-temperature corresponding to a photon wavelength of 0.88 μm, in the infrared. Adding aluminium to form the $Ga_{1-x}Al_xAs$ alloy *increases* the semiconductor band gap, but the band gap changes from direct to indirect at $x = 0.44$. The energy gap at $x = 0.44$ is 1.96 eV, giving a photon wavelength of 633 nm, which is in the red part of the visible spectrum. Tying this altogether means that the AlGaAs system can be used to make LEDs covering the range 880 nm–633 nm, from the near infrared, to just into the red.

I go into the subject of *heterojunctions* in more detail in Chapter 10 but will consider their use briefly here, in a high-brightness red LED structure, as shown in Figure 7.3. Heterojunction means that we have different materials on either side of the p–n junction. If we have the *same* material on either side of the junction, as we do in an ordinary silicon p–n junction diode, the junction is called a *homojunction*. In Figure 7.3 the AlGaAs *composition* is different on the p- and

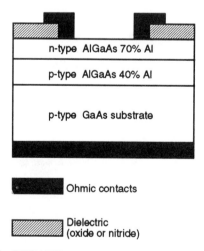

Figure 7.3 AlGaAs/GaAs RED LED.

n-sides of the junction (heterojunction) and this simple change makes a high efficiency (very bright) red LED. The p-AlGaAs layer has only 40% aluminium, so it has a direct band gap. This is the light emitting layer giving photons of wavelength 650 nm. The top n-type layer forms the second half of the junction. This layer has 70% aluminium so it has an indirect band gap, and a *bigger* band gap than the underlying p-layer. These differences have extremely important consequences for the efficiency of the LED. The bigger band gap of the n-layer means that it is a highly efficient injector of electrons into the underlying p-type layer. Carriers like to reside in the material with the *smaller* band gap, so the injected electrons are *confined* to the p-layer due to this difference in band-gap energy. The confinement of the minority carriers to the p-layer pushes up the chances of radiative recombination, and therefore the LED efficiency. Also, because the n-layer has a bigger band gap than the underlying p-layer, the photons produced in the p-layer easily pass through the overlying n-layer. This occurs because the bigger band gap of the n-layer means that it is *transparent to any incident photons*. Finally, the radiative efficiency of p-type material is higher than n-type material, and we have chosen the p-type material to be the photon producer. All these points taken together make the AlGaAs/GaAs LED a *bright*, and very *efficient* source of red light.

The AlGaAs/GaAs LED has given us a photon source that *just* enters the infrared. However, if we want a photon source for fibre optic communications, we need materials that can produce photons of either 1.3 or 1.55 μm wavelength. The reason for needing these particular wavelengths is that conventional optical fibres have *absorption minima* (i.e. maximum transmission, minimum losses) at 1.3 and 1.55 μm. The usual materials system for LEDs (and LASERs) in this wavelength range is GaInAsP (gallium indium arsenide phosphide), which is grown on an indium phosphide crystal (substrate). By changing the composition of the GaInAsP alloy, photon wavelengths between about 1 and 2 μm can be produced.

Materials systems are available that can produce *very* much longer wavelength LEDs. For example, PbSnTe (lead tin telluride) and PbSnSe (lead tin selenide) alloys can emit photons from about 5 μm to near 50 μm!

I've looked briefly at photon emission in the infrared, visible, and ultraviolet parts of the electromagnetic spectrum. I now want to concentrate just on the visible part of the spectrum.

7.3 Materials for visible wavelength LEDs

The visible wavelength LEDs are perhaps the most familiar to us since we see them almost every day, either on calculator displays or indicator panels. You will be most aware of the red LED since it is used extensively as a 'power on' indicator. Yellow, green, and amber LEDs are also widely available, but very few

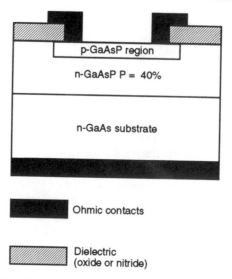

Figure 7.4   GaAsP RED LED on a GaAs substrate.

of you will have seen a blue LED. Why is this? It comes down to basic materials issues that we will examine in this section.

Red LEDs can be made in the GaAsP (gallium arsenide phosphide), GaP (gallium phosphide), and as we saw in the last section, AlGaAs/GaAs systems. $GaAs_{1-x}P_x$ is a direct-gap semiconductor for $0 < x < 0.45$. For $x > 0.45$ the gap goes indirect and for $x = 0.45$ the band-gap energy is 1.98 eV. Hence, the GaAsP compound semiconductor is useful for red LEDs, where its gap remains direct. The GaAsP LED structure (Figure 7.4) may be grown on GaAs crystalline wafers (substrates), which are readily available. The growth technique normally used is chemical vapour deposition (CVD), which you can look up in one of the 'Further Reading' texts listed at the end of this book.

GaP is an *indirect* band-gap semiconductor, and yet we *can* make useful red LEDs using this compound! We are able to do this by one of those quirks of Nature that decides to work with us for a change. The mechanism I am about to describe also works for the GaAsP system where it is used in yellow, amber, and green LEDs. What we can do in some very special cases is introduce an *isoelectronic centre* into an indirect band-gap semiconductor to increase its radiative recombination (photon output) efficiency. Isoelectronic just means that the centre being introduced has the same number of valence electrons as the element it is replacing. For example, nitrogen can replace some of the phosphorus in GaP. It is *isoelectronic* with phosphorus, but behaves quite differently allowing reasonably efficient *green* emission. How do these isoelectronic centres work? Figure 7.5 gives us some idea of what is going on, but don't take this picture too literally. Unfortunately, yet again, we need a result from quantum mechanics.

This time we call upon Heisenberg's Uncertainty Principle, which most of you will have heard of, and which is discussed more fully in Chapter 11. One form of the uncertainty principle tells us that the more precisely a particle's position is defined, the less well-defined is its momentum. For our isoelectronic centre the position is very well-defined (it's on a lattice site), hence there is a considerable spread in its momentum state. I am illustrating this situation in the energy–momentum diagram shown in Figure 7.5. Because the isoelectronic centre has the same valence configuration as the phosphorus it is replacing, *it doesn't act as a dopant*, i.e. it is not introducing any extra electrons or holes. What it does do is provide a 'stepping stone' for electrons in E–k space so that transitions can occur that are radiatively efficient. The recombination event shown has no change in k (momentum), so it behaves like a direct transition. Because the effective transition is occurring between the isoelectronic centre and the valence band edge, the photon that is emitted has a lower energy than the band-gap energy. For the case of GaP:N the energy difference dE, shown in Figure 7.5, is of order 50 meV. This is very useful! Since the photon energy is less than the semiconductor band-gap energy it means that the photon is not absorbed by the semiconductor (the semiconductor is effectively transparent to the photon), and so the photon is easily emitted from the material. This lack of absorption pushes up the efficiency of the diode as a photon source. For emission in the red part of the spectrum using GaP the isoelectronic centre introduced contains zinc (Zn) and oxygen (O). These red LEDs are usually designated GaP:ZnO and they are quite

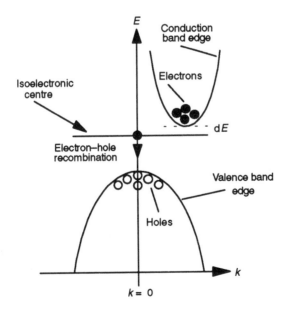

Figure 7.5   Isoelectronic centre.

efficient. Their main drawback is that their emission at 690 nm is in a region where the eye sensitivity is rather low, which means that commercially, the AlGaAs/GaAs diodes are more successful devices.

Orange (620 nm) and *yellow* (590 nm) LEDs are commercially made using the GaAsP system. However, as we have just seen above, the required band-gap energy for emission at these wavelengths means the GaAsP system will have an indirect gap. The isoelectronic centre used in this instance is nitrogen, and the different wavelengths are achieved in these diodes by altering the phosphorus concentration.

Green LEDs (560 nm) have already been discussed. They are manufactured using the GaP system with nitrogen as the isoelectronic centre.

This brings us finally to the short-wavelength end of the visible spectrum; to the blue region. Blue LEDs are commercially available and are fabricated using silicon carbide (SiC). Devices are also made based on gallium nitride (GaN), a material we looked at a little earlier. Unfortunately, both of these materials systems have major drawbacks which render these devices inefficient. The reason silicon carbide has a low efficiency as an LED material is that it has an indirect gap, and no 'magic' isoelectronic centre has been found to date. The transitions that give rise to blue photon emission in SiC are between the bands and *doping centres* in the SiC. The dopants used in manufacturing SiC LEDs are nitrogen for n-type doping, and aluminium for p-type doping. The extreme hardness of SiC makes it a very difficult material to work with! LED fabrication using SiC also requires extremely high processing temperatures.

Gallium nitride (GaN) has the advantage of being a direct-gap semiconductor, but has the major disadvantage that bulk material cannot be made p-type. GaN, as grown, is naturally n^{++}. Light emitting structures are made by producing an intrinsic GaN layer using heavy zinc doping. Light emission occurs when electrons are injected from an n^+ GaN layer into the intrinsic Zn-doped region. A possible device structure is shown in Figure 7.6. Unfortunately, the recombination process that leads to photon production involves the Zn impurity centres, and photon emission processes involving impurity centres are much less efficient than band-to-band processes.

It is *generally* true to say that if we *order* the photon producing processes (in semiconductors) in terms of efficiency, we would get a list like the one below.

(a) band-to-band recombination in direct gap material,
(b) recombination via isoelectronic centres,
(c) recombination via impurity (not isoelectronic) centres,
(d) band-to-band recombination in indirect-gap materials.

So, the current situation is that we *do* have low-efficiency blue LEDs commercially available. We are now awaiting a new materials system, or a breakthrough in GaN (good p-type doping) or SiC (an efficient impurity centre?) technology, for blue LEDs of higher brightness and higher efficiency to be produced.

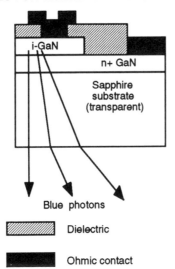

Blue photons

Dielectric

Ohmic contact

Figure 7.6 Blue LED.

Calculation

The following LED data was obtained from an electronics catalogue.

	I_F(typ)/mA	V_F(at I_F typ)/V	Peak wavelength/nm
Red	10	2	635
Green	10	2	562
Yellow	10	2.1	585

where, I_F is the forward current and V_F is the forward voltage dropped across the diode at the given forward current. Assume that the diode construction is $n^{++}p$, so that the current is almost entirely due to the flow of electrons.

For each type of diode calculate,

(a) the band-gap energy of the semiconductor,
(b) the optical power emitted by the LED assuming 30% radiative recombination efficiency (on average three recombination events results in the production of one photon),
(c) the input power to the diode,
(d) the overall electrical-to-optical efficiency.

(a) The band-gap energy is obtained by applying equation (2.2), which gives the following answers:

Band-gap energy (eV)	
Red	1.953
Green	2.206
Yellow	2.120

(b) In each case the current flowing is 10 mA. This corresponds to $10^{-2}/q$ electrons per second where q is the charge on an electron. The energy of a photon is given by $h \times v$ where h is Planck's constant and v is the photon frequency. Notice that hv/q is just the band gap of the semiconductor (in eV) being considered! Hence, the optical power emitted by the LED will be $0.3 \times 10^{-2} \times$ semiconductor band gap in eV. Using this result we obtain

Optical power emitted (mW)	
Red	5.86
Green	6.62
Yellow	6.36

where we have used the factor 0.3 to take into account the 30% efficiency.

(c) The electrical power supplied to the LED is simply $V_F \times I_F$, so we get for each LED

Electrical power supplied (mW)	
Red	20
Green	20
Yellow	21

which if we compare to the results of section (b) gives the following overall efficiencies:

Overall efficiency (%)	
Red	29.3
Green	33.1
Yellow	30.3

Notice a discrepancy here. If the radiative recombination efficiency were 100% then the green LED would seem to give out more light power than electrical power supplied! This comes down to the forward voltage drop (in V) needing to be greater than the band gap energy (in eV) if we are to avoid this problem. Since the catalogue data gives a smaller voltage drop than band-gap energy for the green LED we can expect a problem!

7.4 Junction photodetectors

There is a nice symmetry between LEDs and junction photodetectors. We have seen that we can *produce* photons by *forward* biasing a p–n junction made from certain semiconductor materials. We can *detect* photons by *reverse* biasing a semiconductor p–n junction! However, the symmetry breaks down when we consider the materials for detectors. It is fortunate that for detectors we only need to choose a material with the *right size* energy gap. We don't need a direct-gap semiconductor. This considerably relaxes the problem of trying to find the right photodetector material.

Figure 7.7 shows a simple p–n junction photodetector fabricated using silicon technology, together with the circuit symbol for a photodiode. The structure of the device, as you can see, is quite simple. Photons are incident on the top surface of the diode, and if they have sufficient energy on entering the silicon, they will generate electron–hole pairs. To reduce reflection of the photons from the silicon surface an antireflection coating is often applied. This coating can be a single or multilayer dielectric, with the dielectric layer thickness being chosen to minimize photon reflection *at a given wavelength*. Photons will be able to create electron–hole pairs if their energy is greater than the semiconductor band gap. This criterion can be expressed by an equation we have seen several times before, namely,

$$ENERGY \text{ (in eV)} = 1.24/WAVELENGTH \text{ (in microns)} \qquad (2.2)$$

Remember a micron is 10^{-6} m. In this equation the energy is the semiconductor band gap and the wavelength is the incident photon wavelength. If we put some semiconductor band gaps into this equation we'll find the *cut-off* wavelengths beyond which no electron–hole pairs will be produced (the semiconductor is non-absorbing).

	Band gap (eV)	Cut-off wavelength (μm)
Silicon	1.1	1.13
Germanium	0.67	1.85
Gallium arsenide	1.43	0.87
Gallium phosphide	2.26	0.55
Indium phosphide	1.35	0.92

So silicon will detect photons with wavelengths just getting into the infrared. This means that silicon is a very good detector of photons with slightly shorter

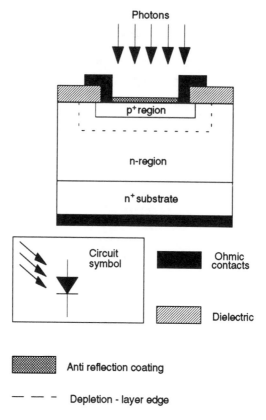

Figure 7.7   Silicon photodetector.

wavelength, i.e. around the red region of the spectrum. You might think that even shorter wavelengths will mean that the silicon will be an even better detector; in fact it gets worse! To see the reason for this we need to start looking at the physics of the detection process.

Figure 7.8 shows the energy-band diagram for a silicon p–n junction operated under reverse bias conditions. As you see, photons are incident on the top surface, and depending on their energy they may penetrate some distance into the diode. Remember, for wavelengths greater than the cut-off wavelength no electron–hole pairs will be produced. The silicon is therefore *transparent* at these wavelengths (neglecting absorption by impurities and defects). For wavelengths slightly shorter than the cut-off wavelength the photons will penetrate some distance into the silicon before absorption takes place, and electron–hole pairs will be produced where the photon is absorbed. The absorption process (if it occurs) is an exponential function of distance into the silicon given by

$$I = I_0 \exp(-\alpha x) \tag{7.1}$$

where

I_0 is the incident photon intensity (watts per unit area),
I is the photon intensity at distance x into the silicon,
α is the absorption coefficient (per unit distance).

If we now consider the mechanisms that give rise to an external current when a photon is absorbed, you'll see why short wavelength photons are not so effective. Returning to Figure 7.8 you'll see some absorbed photons producing electron–hole pairs. There's one in the depletion region and a couple just beyond the edge of the depletion region in the neutral end regions. For the electron–hole pair produced in the depletion region, notice how the charges are very effectively separated by the energy-slope, which is a product of the acting electric field. In fact, the electric field in this region is usually so large that the charges separate at nearly the saturation drift velocity (10^7 cm/s for silicon). This process is therefore rather quick and the speed limitation is set by the width of the depletion region

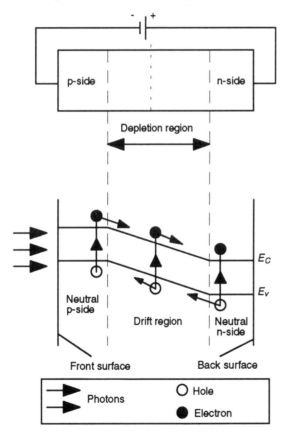

Figure 7.8 Silicon p–n photodetector.

(the distance over which the separated charges must travel). The separated charges will then reach the end ohmic contacts and an external current will flow. Current may also flow if electron–hole pairs are created within a diffusion length or so of the depletion-layer edge. This is the situation being considered just in the neutral regions of the diode. In this case one of the carriers may just diffuse to the edge of the depletion region where it will be rapidly swept across to be detected as an external current. However, the initial diffusion process by which the carrier gets to the edge of the depletion region is rather slow. At this point we should be able to start piecing together what makes a 'good' photodiode, and why short wavelength photons are not very effective.

Photon absorption is a strong function of photon wavelength and for short-wavelength photons the absorption coefficient is very big, i.e. the photons do not penetrate very far into the silicon. This being the case, our buried junction detector shown in Figure 7.8 will not detect short-wavelength photons because they will be absorbed very near the silicon surface. Absorption near the surface means that the electron–hole pairs will recombine *before* they get near the depletion region, charge separation will not occur, and an external current will not flow. To detect these short-wavelength photons you need to make the junction *very* near to the silicon surface. This *can* be done with metal–semi-conductor (Schottky) diodes where a very thin metal layer forms a rectifying contact with the surface of the semiconductor.

Let's now piece together some parameters that will make us a 'good' photodetector. The most effective part of the diode for detection is the depletion region. We would like to make this region wide, to absorb as many of the incoming photons as possible. To make the depletion region wide we operate at a large reverse bias and form one side of the junction in low-doped material. If we want a 'fast' photodiode we want to minimize the 'slow' diffusion process. We can do this by making the top junction contact thin and highly doped. By making the top junction thin we cut down the chances of photon absorption occurring in neutral material, which will lead to (slow) carrier diffusion. By making the top junction highly doped we can make the lower junction in low-doped material, which will give us the wide depletion layer we're after. A wide depletion layer also means a smaller junction *capacitance*, so the diode will operate more quickly because of this as well. In fact, we really only lose out on a wide depletion region because of the increased *transit time* for carriers to cross the depletion region. A wide depletion region will give us a good quantum efficiency (high likelihood of absorbing a photon) and a low junction capacitance, but the penalty is increased carrier transit time. That, then, is the main trade-off that you need to consider in photodiode design.

Calculation

You are to design a high-speed silicon photodiode to receive information from a helium–neon laser (wavelength 632.8 nm) which is modulated at 6 GHz. You are told that the optimum width of the depletion region for this application is that width that gives a depletion-region transit time which is half the modulation frequency. How wide should the depletion width be?

Half the modulation frequency gives us 3 GHz or 3.33×10^{-10} s. If we assume that the carriers are travelling at the saturation drift velocity of about 10^7 cm/s then the depletion-region width is given by

$$WIDTH = 3.33 \times 10^{-10} \times 10^7 \, cm = 33.3 \, \mu m$$

There may be a couple of other free parameters at our disposal, such as the photon wavelength and the detector material, but quite often these are fixed, and so we are stuck with tailoring the depletion-region width. If this is the case, and depletion-layer width is so important, then we'd better be able to define that width in a better way than just using the reverse bias and doping level. Fortunately, we can define the depletion-layer width in a better way. We do this using the p–i–n photodetector.

Figure 7.9 shows the p–i–n photodetector structure and energy-band diagram. The i-region, which stands for intrinsic region, is low-doped (not actually intrinsic). The thin p$^+$ and n$^+$ regions, together with the reverse bias ensures that the i-region is always depleted. The structure is such that virtually all the applied reverse bias appears across the i-region. This produces a large electric field and ensures rapid carrier sweep-out. The quantum efficiency and frequency response are thus optimized by calculating the optimum i-region width in the p–i–n photodiode.

7.5 Photoconductor

The photoconductive detector is *not* a *junction* detector, it's simply a slab of semiconductor with a pair of ohmic end contacts. We've already considered its operation with a calculation in Chapter 4, but the simple photoconductor has a property that warrants a quick study. The photoconductor can exhibit gain! Yes, this simple device, without any junctions, can give more current carriers to an external circuit than incoming photons. I want you to see how it can do this.

Figure 7.10 shows the basic photoconductive cell. The thickness of the cell, d, is such that all incident photons are absorbed. As such, d should be a little greater than $1/\alpha$. The photoconductive material may be either an intrinsic or extrinsic semiconductor (see Figure 7.11), which leads to two different photoconduction mechanisms. Intrinsic photoconduction is illustrated in Figure 7.11(a). The incoming photon must have an energy greater than the semiconductor band gap in

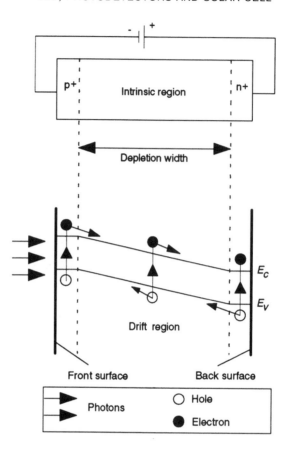

Figure 7.9   p–i–n photodetector.

order to produce an electron–hole pair. The electron–hole pair will then move, under the influence of an external electric field, to the ohmic end contacts. On leaving the contacts a current will flow in the external circuit. The mechanism for extrinsic photoconduction is shown in Figure 7.11(b), (c). In this instance, the carrier which provides current to the external circuit is photoionized from either a donor or an acceptor level. Clearly, the photon energy in these cases is always less than half the semiconductor band-gap energy. That's all there is to the basic mechanism; so how do we get gain from the device?

I shall go through the analysis with an *extrinsic* photoconductor because it's easier to keep track of just one carrier type, but exactly the same argument applies to the intrinsic case. We'll make this analysis a thought experiment with the arrangement shown in Figure 7.12. In this experiment we have a tiny piece of semiconductor with just one donor level in it, and we will photoionize this donor

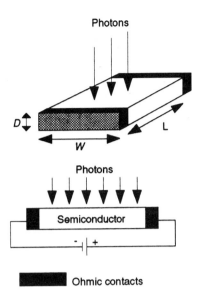

Figure 7.10 Photoconductor.

with just one photon (aren't thought experiments extremely convenient!).
Although there's no real need, we'll also keep things simple by having the
semiconductor sit at a very low temperature, so no current flows (in the dark)
even with an external electric field applied. So, looking at Figure 7.12, in case (a)
the electron sits on its donor level and no current flows (there are no free
carriers). Now we let in one photon, with sufficient energy to ionize the donor, as
in case (b). The electron moves off in the direction shown until it reaches the end
ohmic contact. What happens then? As shown in (b), in order to maintain charge
neutrality (just as we did in the p–n junction analysis) as the electron leaves the
photoconductor, another electron enters the first ohmic contact (refer to Figure
7.11(b) again). So now we've got a current of two electrons for our one incident
photon. Clearly, this can go on even further, but for how long? The process will
come to an end after a time τ (the carrier lifetime) when the donor centre
recaptures the conduction band electron (Figure 7.12(c)). So how much gain do
we get out of such a detector? If the time taken for the electron to traverse the
detector (the carrier transit time) is t_{tr}, and the carrier lifetime is τ then the gain
will be

$$GAIN = \frac{\tau}{t_{tr}} \qquad (7.2)$$

This equation shows us where the main trade-off occurs for the photoconductive
detector. If you want a fast acting device it will require a low lifetime material,
and therefore its sensitivity will be poor (low gain). A high-gain material will be

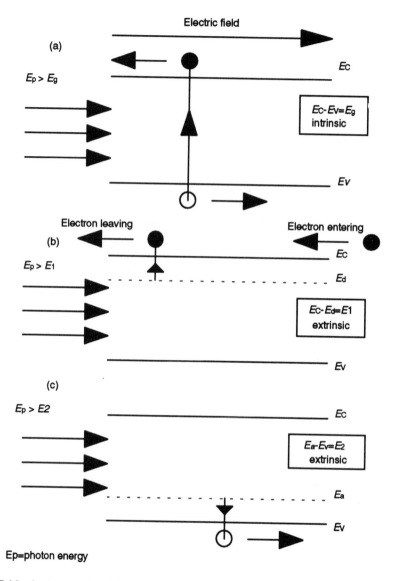

Figure 7.11 Intrinsic and extrinsic photoconductors.

sensitive to low light levels but have poor frequency response. Very high-speed photoconductors have been made in highly lifetime killed silicon using iron as the lifetime killer. These devices can have time responses of better than a picosecond, but naturally they have poor sensitivity.

We have a feel for the physical basis of gain in a photoconductor; now let's have a quick look at the maths.

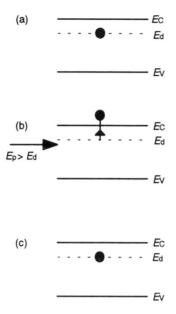

Figure 7.12 Photoconductive gain.

7.6 Photoconductive gain analysis

Consider the photoconductive cell shown in Figure 7.10 again. If we suddenly produce some excess carriers with a light pulse, and then watch how these carriers recombine with time, we get an equation of the form

$$n = n_0 \exp\left(-\frac{t}{\tau}\right) \tag{7.3}$$

In this equation:

n_0 is the concentration of carriers at time $t = 0$ (cm^{-3})
n is the concentration of carriers at time t (cm^{-3})
τ is the carrier lifetime (s).

The recombination *rate* is therefore simply $1/\tau$. If we irradiate our photoconductor with a uniform photon power (with the photon energy slightly greater than the photoconductor band gap), then we can calculate the rate at which we will generate carriers in the photoconductor. If the incident optical power is P_{OPT}, then the *number* of photons uniformly arriving at the surface of the photoconductor *per second* is given by

$$NUMBER\ OF\ PHOTONS\ PER\ SECOND = \frac{P_{OPT}}{h\nu} \tag{7.4}$$

where ν is the incident photon frequency.

Not all of these photons will be effective in producing free carriers. The *quantum efficiency* η gives us the number of free carriers produced per incident photon; hence,

$$\frac{NUMBER\ OF\ FREE\ CARRIERS}{GENERATED\ PER\ SECOND} = \eta\frac{P_{OPT}}{h\nu} \quad (7.5)$$

And the number of free carriers *per unit volume* of photoconductor being *generated per second* will be

$$G = \eta\left(\frac{P_{OPT}/h\nu}{WLD}\right) \quad (7.6)$$

where

G is the free carrier generation rate $(cm^{-3}s^{-1})$,
WLD is the total volume of the photoconductor (cm^3).

The *recombination rate* is given by $R = n/\tau(cm^{-3}s^{-1})$, and in the steady-state we can equate the recombination rate and the generation rate:

$$\frac{n}{\tau} = \eta\left(\frac{P_{OPT}/h\nu}{WLD}\right) \quad (7.7)$$

Let's now write down the photocurrent, I_p, that is flowing:

$$I_p = E\sigma DW = DWqn\nu_d \quad (7.8)$$

where

$\mathbf{E}$ is the acting electric field $(V\,cm^{-1})$,
σ is the semiconductor conductivity $(S\,cm^{-1})$,
ν_d is the carrier drift velocity $(cm\,s^{-1})$.

If we substitute the carrier concentration, n, from (7.7) into (7.8) we get

$$I_p = q\tau\eta\left(\frac{P_{OPT}}{h\nu}\right)\left(\frac{\nu_d}{L}\right) \quad (7.9)$$

The primary photocurrent, I_{PRIM}, is defined as the current flowing due to the incident *effective* photons (i.e. taking the quantum efficiency into account). We can therefore write for the *primary* photocurrent

$$I_{PRIM} = q\eta\left(\frac{P_{OPT}}{h\nu}\right) \quad (7.10)$$

Finally, we calculate the photoconductive gain by dividing the photocurrent, equation (7.9), by the primary photocurrent, equation (7.10):

$$GAIN = \frac{I_p}{I_{PRIM}} = \frac{\nu_d\tau}{L} \quad (7.11)$$

But the carrier transit time t_{tr} is L/ν_d, so (7.11) can be rewritten as

$$GAIN = \frac{\tau}{t_{tr}} \quad (7.12)$$

which is fortunately the same equation we obtained earlier using the thought experiment approach.

7.7 Solar-cell

We have seen how to generate light from a p–n junction and how to detect photons using a p–n junction. We're now going to see how to generate power from light using a p–n junction. The humble p–n junction is an amazingly versatile device!

In Section 7.4 we used a *reverse biased* p–n junction as a photodetector. In actual fact we didn't have to reverse bias the junction at all for its use as a detector. If we illuminate a p–n junction, with no external bias, we still get the same charge separation mechanism operating in (and close to) the depletion region. These moving charges can be detected as an external photocurrent or we could measure the photovoltage produced at the diode terminals. The *I–V* characteristics of a p–n junction diode under different levels of illumination (unless stated otherwise I will assume the light energy is greater than band gap) are shown in Figure 7.13. Notice that the illuminated junction will produce a terminal voltage at zero current. This is the open circuit voltage V_{OC}, and this effect is known as the *photovoltaic effect*. We can calculate the magnitude of this voltage using the ideal diode equation (6.57):

$$I = I_0 \left(\exp\left[\frac{qV}{kT}\right] - 1 \right) \quad \text{IDEAL DIODE EQUATION} \qquad (6.57)$$

If equation (6.57) is rearranged to make V the subject we get

$$V = \frac{kT}{q} \ln\left(1 + \frac{I}{I_0}\right) \qquad (7.13)$$

If we substitute the photocurrent for I in (7.13) then we will obtain the open-circuit photovoltage:

$$V_{OC} = \frac{kT}{q} \ln\left(1 + \frac{I_p}{I_0}\right) \qquad (7.14)$$

We can estimate the maximum value of V_{OC} for p–n junctions in the following way. The open-circuit voltage cannot be larger than the built-in junction potential that provides it, and the junction potential is usually less than the band-gap voltage (refer back to Figure 6.8 to see why). So we can generally write

$$V_{OC} < V_{bi} < \frac{E_g}{q} \qquad (7.15)$$

This being the case, silicon solar-cell open-circuit voltages are typically under 1 volt. A good silicon solar-cell will provide a short-circuit current of about 20 mA or more per square centimetre of diode area. These values are not particularly staggering if we want to produce useful power with solar-cells. This means we have to put many cells in series to push up the voltage, and many cells in parallel to push up the current. In practice this makes the silicon solar-cell a *large area device*, often using a whole (4- or 6-inch diameter) wafer for its fabrication.

Before briefly discussing the structure of silicon solar-cells I will mention one

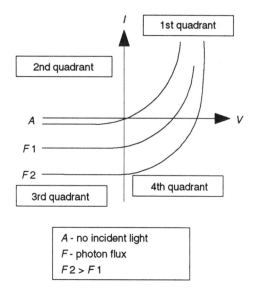

Figure 7.13   *I–V* characteristic of illuminated p–n junction.

useful figure-of-merit called the *fill-factor*. Figure 7.14 shows the solar-cell *I–V* characteristic for just one level of illumination. As before, V_{OC} is the open-circuit voltage and I_{SC} is the short-circuit current developed at this level of illumination. However, the maximum power that can be supplied under these conditions is shown by the *maximum power rectangle* in Figure 7.14. I_m and V_m correspond to the maximum current and voltage that the cell can supply. The figure-of-merit, called the fill-factor, is given by

$$FILL\ FACTOR = \frac{I_m V_m}{I_{sc} V_{OC}} \tag{7.16}$$

which geometrically is the area of the maximum power rectangle divided by the area the *I–V* curve forms with the current and voltage axes.

My final discussion in this chapter will be on some aspects of solar-cell design. I will only briefly look at some aspects because solar-cell design could fill a book on its own. Recall the trade-offs we met in p–n junction photodetector design. Well, solar-cell design is several orders of magnitude worse!

With a conventional photodiode it is very likely that we only want to detect one particular wavelength of light. That fact alone makes several aspects of photodiode design pretty straightforward. Now look at the solar-cell case. The solar spectrum is not a monochromatic source! The Sun provides reasonable power anywhere between about 0.3 and 1.5 microns. Our silicon solar-cell is not going to be effective for all the radiation with wavelength beyond 1 micron (it's transparent), and as we saw earlier it's pretty inefficient for the short-wavelength

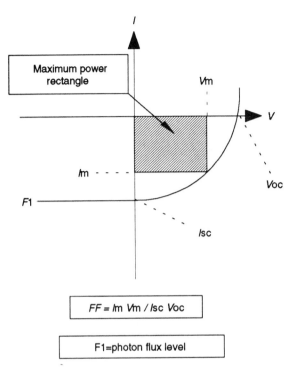

Figure 7.14 Fill factor.

end as well. This is not a good start. One problem of course is that we're trying to detect a broadband radiation source with a semiconductor with a single-value of energy gap. We *could* be clever and use several different semiconductors in our cell to try and cover more of the spectrum, but this introduces more problems than it solves. For one, it is very difficult just from a materials point of view to make a cell from several different semiconductors. Then, if you're making a cell containing a couple of p–n junctions, the two cells have to be well matched electrically or efficiency will suffer. In spite of all the difficulties efficient multijunction solar cells have been fabricated, but usually in compound semiconductors rather than silicon.

We actually need to get the sunlight *into* the silicon before it can be used. Obviously, an antireflection coating helps us a lot, but again these are usually only efficient for *one particular wavelength*. Solar-cell fabricators are so keen to get as high an efficiency as possible that they go to great lengths to get as much useful solar radiation into the cell as possible. One approach uses the preferential wet-etching of certain atomic planes in silicon. The outcome of this procedure is to cover the surface of the solar-cell with lots of inverted pyramidal dimples as in Figure 7.15. These tiny pyramids, less than 100 microns across, help to get more light *into* the cell, as shown in the diagram.

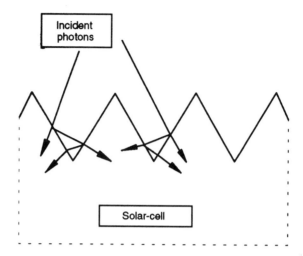

Figure 7.15   Solar-cell with pyramidal top surface.

Finally, we need to get the photocurrent out of the cell, and this will need both front and back contacts. The back contacts are not a great problem, but the front contacts are going to get in the way of the incoming solar radiation! Things are made worse by the fact that series resistance in the solar-cell *seriously* affects cell efficiency. We would actually like a nice thick metallic conductor all over the front surface of the cell, but then of course no light would get in. A compromise is met by having narrow metallic fingers across the cell front surface that link up to a thick metallic 'bus bar' at the edge of the cell. Making good low-resistance metal contacts on the solar-cell is a non-trivial problem!

I have barely touched on the practical problems of solar-cell fabrication and yet things are already looking pretty bad. It is therefore amazing that cell efficiencies in excess of 25% have been achieved in laboratory devices where the maximum theoretical efficiency is just over 30%. This amply illustrates the ingenuity of the solar-cell designers in attacking these problems!

CHAPTER EIGHT

Bipolar transistor

For the first time in this book we encounter a bipolar (two-carrier) device with *three* terminals. I can clearly recall the problems I had trying to understand how a device could possibly have *three* terminals. Two is easy, it's just in and out. But three! What's the third one for? Because of these conceptual problems I will break from the standard physical electronics format in this chapter, and, apart from discussing the physics of the bipolar transistor, I will show how they are used in a couple of *simple* circuits. The circuits I will give are there to be built (if you haven't done this sort of thing before). If you build the circuits as shown, you should not damage any devices. If you *do* manage to destroy a transistor don't panic, and please don't give up.

As an attempt at motivating you to play with electronics, let me tell you how I started practical electronics from a near-zero knowledge-base. I bought some old computer boards in the early 1970s, removed some transistors and put them into a simple circuit with a torch bulb, see Figure 8.1. Note no resistors, and certainly nothing as complicated as a capacitor. With my near-zero knowledge-base I noted that the bulb shone brightly when the base connection was put on the positive battery terminal, but didn't shine at all when disconnected or put on the negative terminal. The other thing that happened was that when the bulb was *on*, the transistor got too hot to touch and didn't work any more after a couple of seconds. After destroying maybe ten transistors I started to get the message. I went away and got a simple magazine on hobby electronics and started to learn what I was doing wrong. It worked for me but I don't expect this to be a universally applicable learning technique. I hope that you will understand the basics of the bipolar transistor, and can build a couple of working circuits by the end of this chapter without destroying *any* transistors.

Before launching into this device I'd like to make a couple of comments that I feel are important. A significant number of the 'new wave' of students coming to be taught electronics seem to think the transistor (meaning the bipolar transistor throughout this chapter) was *discovered*, almost as if by luck. Bardeen, Brattain and Shockley *did not discover* the transistor! The work being carried out (at Bell Labs. in the USA) was a well-thought out, intense programme, trying to

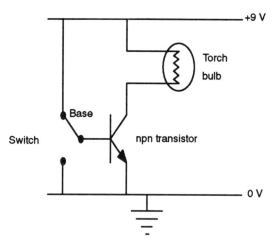

Figure 8.1   Transistor destroyer.

understand the mechanisms of solid-state physics at the atomic level. The scientists mentioned above were experts in the field of solid-state physics, and could apply that knowledge to eventually *evolve* the bipolar transistor (I don't even like the word *invent* here, although it is probably applicable). The point I am trying to make is that, true enough, the transistor's time had come, but it was due to the accumulation of a wealth of knowledge, not the fortuitous throwing together of semiconductors and dopants. It is interesting to note that most of the *basic theory* of the new device was finished within a couple of years of its fabrication, around 1950. This brings me to my second point. I have seen in some 'standard' texts that the bipolar transistor is a *simple* device. It's funny how a simple device, the workings of which have been known since 1950, still commands so much of our time and effort trying to improve it!! The *structure* of the bipolar transistor may *appear* simple, but even that is an illusion. This is one of those cases I mentioned earlier where ostensibly simple systems can show extremely complex behaviour. That's why (and it amazes me as well) we are still working on this fundamental device and learning more about it every year. In the latter part of this chapter I will take you to the forefronts of the new bipolar transistor research to show you what's going on. I want you to realize that now we can get right inside the transistor and start playing around with its fundamental properties (tailoring I believe is the 'in' term). It's almost like having a completely new device to deal with. The transistor *is not* simple. I stated in Chapter 6 that the p–n junction was not simple. Well the transistor is composed of *two* p–n junctions, and they can *interact* with each other as well!!

8.1 Basic concepts

The bipolar transistor is a three-terminal, three-layer device. It comes in two versions, pnp, and npn. The npn version is the more common of the two and is shown in Figure 8.2. The pnp structure, and circuit symbol, is illustrated in Figure 8.3. You will see in many of the 'standard' texts that the pnp transistor is usually the type considered in detail. The reason for this is that, as you know, *conventional current* arrows point in the direction in which *holes* would flow. In the *pnp* transistor the *hole* currents are the *primary* currents, and so the arrows point in the direction in which the primary currents are flowing. This is meant to make things easier for the student. Since the npn transistor is the more common of the two, this is the one *I* will consider in more detail. I am assuming that students reading *this* text have sufficient intelligence that they're not put off by a current arrow pointing in the *opposite* direction to electron flow!

One of the major problems I had in getting to grips with the bipolar transistor was the *terminology* used for its modes of operation. I *still* think this terminology is confusing, but as it is standard, I will use the *same terms* but I will explain their meaning in more detail than usual.

As the bipolar transistor has three terminals it's not too surprising that there are three different ways of connecting it up. Look at Figure 8.4, which shows the three different connections. Notice that one of the terminals is *common* to input

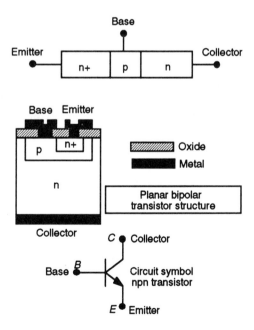

Figure 8.2 npn transistor.

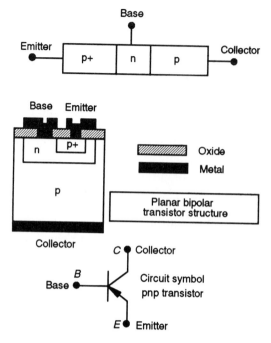

Figure 8.3   pnp transistor.

and output in each configuration (this gives the *name* to the different connection configurations). This, I feel, starts to make the whole situation less confusing with regard to the use of the third terminal. With a two-terminal device, such as a p–n junction diode or Gunn diode, it can be difficult to electrically isolate the output side from the input side. This isolation is more easily achieved in three-terminal devices (there's an extra junction to play with) and you have some external control over the device via the third terminal (which is missing altogether in the diode).

Let's get straight into how the bipolar transistor works by looking at its normal mode of operation, called the *active mode*. Figure 8.5 shows an npn transistor operating in the active mode. The emitter–base junction is *forward biased* and the collector–base junction is *reverse biased*. Note that if the base region is *thin enough*, electrons injected into the p-type base from the forward biased emitter–base junction (diode) have got a chance of *diffusing* across the base to get to the collector–base junction. Why *diffusing*? Because away from the emitter–base and collector–base depletion regions the base is neutral. This neutral base region (assuming it's uniformly doped) doesn't have an electric field across it, therefore any carrier movement is by diffusion. How thin should the base be for the electrons to get across? Well, if the base width is a fraction (say 1/10th) of the electron *diffusion length*, then most of the electrons are likely to get across before

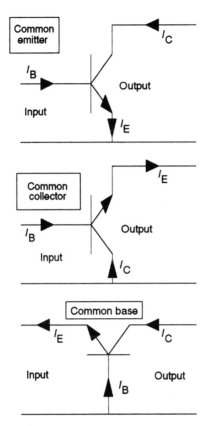

Figure 8.4   npn circuit connections.

recombining with the majority carrier holes in the base. What happens to these electrons when they do get across the base? Remember the collector–base junction is reverse-biased. Electrons reaching the edge of the collector–base junction depletion region find themselves in an electric field which accelerates them into the collector! Think a little bit about this situation and it makes perfect sense from our p–n junction knowledge. The collector–base junction is reverse-biased. Electrons appearing near the depletion-layer edge in the base are *minority carriers*. The reverse leakage current in the p–n junction is due to these minority carriers. Hence, if we deliberately introduce loads of minority carriers next to the depletion region of a reverse-biased junction we're going to see a massive increase in the reverse current. Let's briefly recap what's already been said. Electrons are injected from the emitter into the base. The electrons then diffuse across the base and are collected (hence the name) by the collector. If the base region is thin enough, virtually no electrons recombine with the majority carrier holes, so the *injected* electron current equals the *collected* electron current. So far this is all

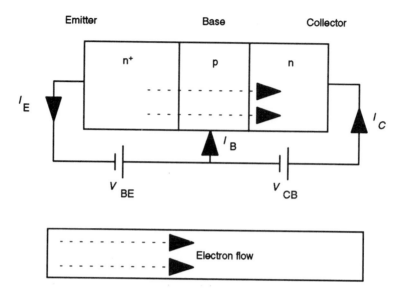

Figure 8.5   npn transistor active mode.

pretty straightforward, if not dull, so what's all the fuss about with these transistors? We haven't looked at the base current, I_B, yet!

Refer back to Figure 8.5 again. What is this base current I_B? I'm assuming this is a 'good' transistor with a narrow base and no electron–hole recombination in the base (all the injected electrons get across the base). So the base current *is not* a recombination current. If it's not a recombination current, then it's obvious what this current is! The emitter–base junction is *forward biased*, electrons are injected from the emitter into the base, and *holes are injected from the base into the emitter*! The base current is no more than the forward current through the base–emitter junction due to the injected holes. Well so what. That's even more boring. This doesn't seem like a very clever device at all. At this point you've got to get your brain into *reverse* to see why this device is so remarkable. A lot of the basic understanding of the bipolar transistor comes down to knowing what's going on at this emitter–base junction in detail; forget about the collector–base junction just for a moment. The forward biased emitter–base junction will inject lots of electrons into the base, which will be collected by the collector, and it will inject lots of holes into the emitter which gives rise to the (hole) base current. From our hard work in Chapter 6, and referring to equations such as (6.45), you should be aware that the injected minority carrier density is related to the doping density of the injecting contact (via a Boltzmann factor). In English this means the higher-doped side of the junction injects the most carriers, which is logical. *If* we had a p^+n diode, most of the total diode current under forward bias conditions would be carried by *holes*! If we had a forward biased n^+p diode, then most of the total diode current would be carried by electrons. Interesting, but still unimpressive. Keep with it just for this last bit.

Assume for the moment that our npn transistor has an n^+p emitter–base junction. Most of the current flowing through this junction under forward bias conditions is going to be electron current. A much smaller hole current is also flowing. What happens if we cut this hole current off? We could simply put a switch in the base connection in Figure 8.5, and open the circuit. Obviously, the emitter–base diode current stops as well; hardly surprising, we haven't got a forward biased diode any more. And in that simple statement is the key! We can control the *much bigger* electron current flowing from the emitter to the collector using the *tiny* base current! Without the tiny base current the big emitter–collector current does not flow. If we 'jiggled' the tiny base current, we'd get a much bigger 'jiggle' in the collector current. We've got an *amplifier*! If we cut off the tiny base current, we cut off the much bigger collector current. We've got a *switch*! We're off!!

Calculation

We've only just started bipolar transistors and already here's a calculation. You'll see that all you need is the Chapter 4 work to crack it.

An npn transistor with emitter–base and collector–base areas of $0.1\,\text{mm}^2$ has an excess electron density of $8.5 \times 10^{15}\,\text{cm}^{-3}$ maintained at the emitter–base junction. If the base width is $5 \times 10^{-6}\,\text{m}$ and the electron mobility is $1500\,\text{cm}^2/\text{Vs}$ sketch the approximate distribution of electrons in the base and estimate the collector current.

This may seem like a pretty mean one at this stage. I've deliberately mixed up units, but let's attack this problem systematically and watch it crumble!

Figure 8.6 gives the physical picture of the thing we're trying to solve. The npn transistor is shown at the top with the emitter, base, and collector labelled. We're told that the excess electron concentration at the emitter–base junction is $8.5 \times 10^{15}\,\text{cm}^{-3}$ and this is shown in the diagram below the transistor. We're told that the base width is $5 \times 10^{-6}\,\text{m}$ and this is also shown. I've shown the excess electron concentration dropping off to zero at the collector–base junction because I'm assuming that the collector efficiently removes (collects) all the electrons that reach it. I've also shown two possible curves for the excess electron concentration. The solid curve falls off exponentially, implying that the base width is several diffusion lengths long; the dotted linear curve implies the base width is less than a diffusion length. Transistor base widths are usually narrow (for speed) and we shall therefore assume the linear excess electron fall-off with distance. We now need to try and find the electron current with the given information. The current is going to be a diffusion current (we're talking about excess carrier densities, and have just seen that there are no electric fields in a uniformly doped neutral base), so we need to look at the diffusion equations. We've only got the mobility as a useful parameter so it's clear we've got to use the *Einstein relation* to get a diffusion coefficient. With the given mobility we get

$$D_n = \mu_n \frac{kT}{q} = 0.15 \times 25 \times 10^{-3} = 3.75 \times 10^{-3}\,\mathrm{m^2\,s^{-1}}$$

The equation we're going to have to use to find the *diffusion current density* is just (4.44):

$$J_n = qD_n \frac{d\delta n}{dx}$$

We can calculate $d\delta n/dx$; it's just

$$\frac{d\delta n}{dx} = \frac{8.5 \times 10^{21}}{5 \times 10^{-6}} = 1.7 \times 10^{27}\,\mathrm{m^{-4}}$$

Substitute into equation (4.44) to get

$$J_n = 1.6 \times 10^{-19} \times 3.75 \times 10^{-3} \times 1.7 \times 10^{27} = 1.02 \times 10^6\,\mathrm{A\,m^{-2}}$$

Finally, the device area is $0.1\,\mathrm{mm^2}$, so the current is

$$I_n = 0.1 \times 10^{-6} \times 1.02 \times 10^6 = 0.102\,\mathrm{A}$$

Not so difficult really!

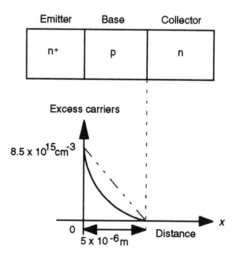

Figure 8.6 Diagram for calculation.

Calculation

We can define the direct-current (d.c.) current gain of a bipolar transistor (in the common-emitter configuration), β, as

$$\beta = \frac{I_C}{I_B} = \frac{COLLECTOR\ CURRENT}{BASE\ CURRENT}$$

We know from the last section that this ratio is roughly going to be the emitter doping divided by the base doping. Why? Because if we have no base recombination, the collector current is equal to the emitter current and the emitter current is proportional to the emitter doping. The base current is proportional to the base doping, so the current gain is proportional to the emitter doping divided by the base doping. Let's be more specific in finding out what β actually is and calculate it for the following npn transistor:

emitter doping N_D $10^{20}\,\mathrm{cm}^{-3}$,
base doping N_A $5 \times 10^{16}\,\mathrm{cm}^{-3}$,
base width W_B $0.8\,\mu\mathrm{m}$,
hole diffusion length L_p $0.35\,\mu\mathrm{m}$ (in $10^{20}\,\mathrm{cm}^{-3}$ doped Si),
hole diffusion coefficient D_p $1.15 \times 10^{-3}\,\mathrm{m}^2\mathrm{s}^{-1}$,
electron diffusion coefficient D_n $3.75 \times 10^{-3}\,\mathrm{m}^2\mathrm{s}^{-1}$

where we calculated the hole and electron diffusion coefficients in Section 4.10. Assume that the emitter width is much greater than the minority carrier diffusion length.

Our next step is to find equations for the electron and hole diffusion current densities flowing in a forward biased p–n junction. We have already derived these equations in Chapter 6. The equation for electrons is

$$J_n = -\frac{qD_n n_{p0}}{L_n}\left(\exp\left[\frac{qV}{kT}\right] - 1\right) \tag{6.53}$$

which we must modify in this instance because the characteristic distance over which the electrons are diffusing in the transistor is *not* the electron diffusion length, L_n, but the *much smaller* transistor base width W_B. This change means we must use the equation

$$J_n = -\frac{qD_n n_{p0}}{W_B}\left(\exp\left[\frac{qV}{kT}\right] - 1\right) \tag{8.1}$$

The equation for the hole current density is given by

$$J_p = \frac{qD_p p_{n0}}{L_p}\left(\exp\left[\frac{qV}{kT}\right] - 1\right) \tag{6.54}$$

For the magnitude of the current gain we simply divide equation (8.1) by equation (6.54) (neglecting signs) to get

$$\beta = \frac{D_n n_{p0} L_p}{D_p p_{n0} W_B} \tag{8.2}$$

We're at the last hurdle, but this is the tricky bit. We've got to get rid of the n_{p0} and p_{n0} for quantities we know, namely N_D(emitter) and N_A(base). This is easy *provided* you keep track of what's the base and what's the emitter. You'll see why now. We'll use the equations given in (6.37), namely,

$$\left. \begin{array}{l} n_{n0} \cdot p_{n0} = n_i^2 \quad \text{n-type material (the emitter)} \\[2ex] p_{p0} \cdot n_{p0} = n_i^2 \quad \text{p-type material (the base)} \end{array} \right\} \tag{6.37}$$

Since n_{n0} is N_D the emitter doping, and p_{p0} is N_A the base doping, we can rewrite equation (8.2) as

$$\beta = \frac{D_n N_D L_p}{D_p N_A W_B} \tag{8.3}$$

Finally, substitute the given values for the quantities in equation (8.3):

$$\beta = \frac{37.5 \times 10^{20} \times 0.35 \times 10^{-4}}{11.5 \times 5 \times 10^{16} \times 0.8 \times 10^{-4}} = 2853$$

Notice that a straight doping ratio would have given us a gain of 2000, which is pretty close to the value calculated above.

8.2 Basic structure

In the last section I said the base region had to be *thin* for transistor action. Let's make the base thick, say ten diffusion lengths wide, and see what happens. Figure 8.7 shows such a device. For such a wide base any electrons injected by the emitter recombine with the majority carrier holes in the base well before they can reach the collector. The only current flowing in the collector–base circuit under these conditions is the reverse saturation, or leakage current I_{CBO}. This nomenclature is straightforward. I_{CBO} tells us it's the collector–base current flowing with the emitter *open-circuit* (that's what the 'O' is for). Under these conditions the two well-separated junctions act like two *back-to-back* diodes, also shown in Figure 8.7. For such a wide base there is no gain, and *no transistor action*.

Let's return to the structure of a 'good' transistor. In a 'good' transistor there is very little recombination in the base, hence,

(1) *The base is made thin.*

A thin base will also decrease the base transit time (the time for the injected minority carriers to get across the base).

We want the *emitter* to inject lots of carriers into the base, so:

(2) *The emitter is heavily doped.*

Any excess capacitance at junctions will slow the transistor down when switching. In order to reduce the capacitance at the collector–base junction,

(3) *The collector is lightly doped.*

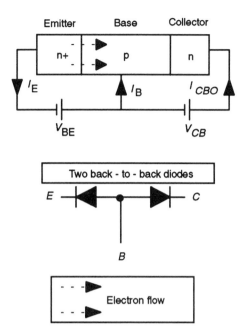

Figure 8.7 Wide base device.

For reasonable values of the current gain, β, we want little injection of majority carriers from the base into the emitter. For an npn transistor this means we want to reduce the hole current into the emitter as much as possible.

(4) *The base is doped several orders of magnitude below the emitter.*

Although the above are good guidelines, they only 'scratch the surface' of the design problems. For example, in point (4), we can't just keep decreasing the base doping, because among other things the *base resistance* will increase. An increasing base resistance will slow the transistor down. We *can* get around this problem, however, by using heterojunctions, or polyemitters, which we will look at later in this chapter.

8.3 Diffusion capacitance

In the above list of guidelines it was said that the collector should be low-doped to lower the collector–base *depletion* capacitance (we looked at depletion capacitance in the p–n junction chapter). There's nothing wrong with that statement, but there is another capacitance to consider whose effect is often many times greater than the depletion capacitance. This is the *diffusion capacitance* associated with the injection of minority carriers. Let's consider the forward biased emitter–base junction for a moment. Under forward bias a lot of (minority carrier) electrons will be injected into the (p-type) base. If we now want to turn the transistor off quickly we've got to get rid of all that stored (minority carrier) electron charge in the base region (which will continue to flow until its gone). This *stored minority charge* is responsible for the diffusion capacitance, therefore the diffusion capacitance will be a function of the *minority carrier lifetime*.

Figure 8.8 shows the *I–V* characteristic of the emitter–base junction (it is of course just a p–n junction diode characteristic) and two equivalent circuits. The two equivalent circuits correspond to the two points, A and B, on the *I–V* curve. For point A on the *I–V* curve we are in reverse bias so the capacitance is just the depletion capacitance, and the input resistance will be *large*. For point B on the *I–V* curve we are in forward bias so the capacitance will be *both* the depletion capacitance and the diffusion capacitance combined. Also, the emitter (input) resistance will be *low*.

We can estimate the emitter resistance under forward bias conditions in the following way. Start with the ideal diode equation with forward bias V_F applied:

$$I = I_0 \left(\exp\left[\frac{qV_F}{kT} \right] - 1 \right) \tag{6.57}$$

Since $V = IR$ (Ohm's law) we can differentiate equation (6.57) to find the reciprocal emitter resistance:

$$\frac{1}{r_e} = \frac{dI}{dV_F} = \frac{q}{kT} I_0 \exp\left[\frac{qV_F}{kT} \right] \tag{8.4}$$

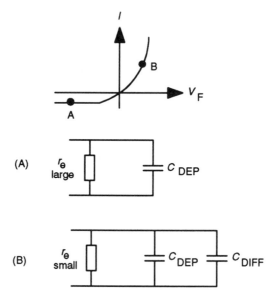

Figure 8.8 Diffusion capacitance.

For large forward bias, the forward current is I_F giving

$$\frac{1}{r_e} = \frac{qI_F}{kT} \qquad (8.5)$$

We already know that $kT/q = 25\,\text{mV}$, so that if we have a forward current of, say, $1\,\text{mA}$ flowing, the emitter resistance will be

$$r_e = 25\,\text{mV}/1\,\text{mA} = 25\,\Omega \qquad (8.6)$$

Notice that device size does not come into the emitter resistance equation. Consequently, the equation for r_e holds for all transistors – it is independent of geometry.

Let's now look at the diffusion capacitance. Recall that we're expecting the minority carrier lifetime to make an appearance here. First we need the equation relating stored charge, voltage and capacitance:

$$Q = CV \qquad (6.59)$$

For our forward biased emitter–base junction, the stored charge is going to depend on the forward bias. From equation (6.59) we can deduce

$$C_{\text{DIFF}} = \frac{d\Delta Q}{dV_F} \qquad (8.7)$$

where

ΔQ is the excess (minority carrier) stored charge injected into the base. I am not considering the (much smaller) injected hole charge into the emitter, although precisely the same equations hold for this case as well.

C_{DIFF} is the diffusion capacitance.

The excess minority carriers recombine at a rate defined by the minority carrier lifetime. A rate of change of charge with time is simply a *current*, so we can write

$$\frac{\Delta Q}{\tau} = I_F \qquad (8.8)$$

where

I_F is the forward current corresponding to the forward bias V_F.
τ is the minority carrier lifetime (s).

Combining (8.7) and (8.8) gives us

$$C_{\text{DIFF}} = \frac{d\Delta Q}{dV_F} = \tau\frac{qI_F}{kT} \qquad (8.9)$$

Equation (8.9) shows how the diffusion capacitance is a function of the forward current and the minority carrier lifetime.

Calculation

Calculate the emitter–base junction diffusion (C_{DIFF}) and depletion capacitance (C_{DEP}) for a bipolar transistor using the following data:

(a) collector current 0.5 mA,
(b) current gain β 150,
(c) depletion-layer width 0.1 μm,
(d) junction area 225 μm^2,
(e) minority carrier lifetime 1 μs.

Remember from equation (6.60) that the depletion capacitance has exactly the same *form* as the parallel plate capacitance. Hence,

$$C_{DEP} = \frac{A\varepsilon_{Si}}{x_{dep}}$$

where

A is the junction area,
x_{dep} is the depletion-layer width,
ε_{Si} is the permittivity of silicon.

Substituting values gives

$$C_{DEP} = \frac{2.25 \times 10^{-10} \times 8.854 \times 10^{-12} \times 11.7}{10^{-7}} = 0.233\,pF$$

For the diffusion capacitance, we start with equation (8.7) using V_{BE} (the base–emitter voltage) for the forward voltage bias. Hence,

$$C_{DIFF} = \frac{d\Delta Q}{dV_{BE}}$$

We know from (8.8) that $\Delta Q = I_B\tau$, since the current that is flowing (due to the recombination of the injected excess electrons with the majority carrier holes) is the base current. Hence,

$$C_{DIFF} = \frac{d(I_B\tau)}{dV_{BE}}$$

We now need to replace the base current (which is not given) by the collector current (which we know). We use the current gain β to do this since $\beta = I_C/I_B$. Finally,

$$C_{DIFF} = \frac{\tau q I_B}{kT} = \frac{\tau q I_C}{kT\beta} = \frac{10^{-6} \times 0.5 \times 10^{-3}}{25 \times 10^{-3} = 150} = 133\,pF$$

So, in this instance, the diffusion capacitance is considerably greater than the junction depletion capacitance.

8.4   Current components

In this section we look in a little more detail at the currents flowing in a bipolar junction transistor (BJT), and see how these currents are related to some commonly used transistor parameters.

Figure 8.9 shows the energy-band diagram for an npn bipolar transistor in thermal equilibrium (a), and in the active mode (b). By seeing how currents will flow in this energy diagram (active mode) we can form a picture of the current components flowing in the npn transistor itself.

Looking at Figure 8.9(b) we see that the emitter–base junction barrier height is lower than in the thermal equilibrium case because the emitter–base junction is forward biased. We know from our p–n junction theory that this forward biased junction will lead to the injection of electrons into the base, and holes into the emitter, as shown. The electrons injected into the base will diffuse across the neutral base region (provided it is thin enough) until they come to the collector–base junction region. Here the electrons see a large potential well caused by the reverse biased collector–base junction, and the electrons 'fall' into the well. Remember, under drift conditions holes 'float' and electrons 'sink'. Notice also that an injected electron has recombined with a hole in the base which will give rise to some base current. These are the major currents that we must consider. There are also some minor currents flowing, such as the reverse leakage hole current I_{Cp} at the collector–base junction. Let's quickly try and transfer these currents from an energy-band diagram to a three-terminal device picture.

Figure 8.10 shows the current components we've just discussed, flowing in an npn bipolar transistor operating in the active mode. Apologies for the nomenclature in the diagram, but there are quite a few currents flowing, and we want to be able to discuss them separately. Let's quickly run through each of

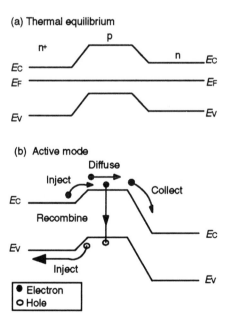

Figure 8.9   Transistor energy-band diagrams.

these current components. I_{En} is the total electron current emitted by the emitter. I_{Cn} is the fraction of the emitter current that is collected by the collector. The difference between the emitted electron current and the collected electron current $I_{En} - I_{Cn}$ is the *recombination* current flowing in the base. I_{Ep} is that component of the total base current due to the *injection* of holes from the base into the emitter. I_{Cp} is the *hole* component of the collector–base reverse *leakage* current. Be very careful about what the arrows are implying in this diagram. The arrows *inside* the transistor are showing the direction in which the electrons and holes are *flowing*. This means the electron *current* is in the *opposite* direction to the electron *flow* (arrows) *inside* the transistor. The arrows *on* the transistor *terminals outside* the transistor show the direction in which the conventional base, emitter, and collector currents flow. We are now at a stage where we can look at some transistor parameters and consider what physical conditions influence them.

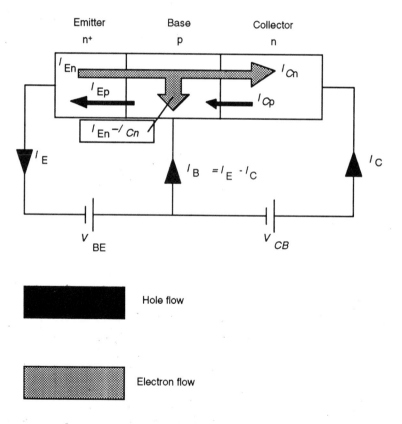

Figure 8.10 Transistor current components.

8.5 BJT parameters

In this section we shall study briefly some of the parameters encountered when considering current flow in the BJT. As you proceed with this section you'll see that you already have a qualitative understanding of most of the concepts. The equations we shall derive will put this understanding on a more quantitative footing.

Return to Figure 8.10. It has been stated that this transistor is operating in the *active mode*. If we look at the base connection, we see that it is common to both the input (between the emitter and the base) and the output (between the collector and the base). The transistor is thus connected in the *common-base configuration*. Let's derive some parameters for the transistor in this configuration.

We know that the collector current is *proportional* to the emitter current, *and* that the collector current includes a reverse leakage component, I_{CBO}. We can write an equation describing this:

$$I_C = \alpha_0 I_E + I_{CBO} \tag{8.10}$$

where α_0 is the constant of proportionality. This constant is given a name, it's called the *common-base current gain*. In fact, as we'll only be considering small current changes (so that we'll be able to take differentials), in this particular case it's called the *small-signal, common-base current gain* (what a mouthful!). If we now differentiate equation (8.10) keeping V_{CB} constant, we'll get rid of the (constant) effect of I_{CBO},

$$\alpha_0 = \left.\frac{\partial I_C}{\partial I_E}\right|_{V_{CB}} \quad \text{SMALL-SIGNAL COMMON-BASE CURRENT GAIN} \quad (8.11)$$

You may also see α_0 called the *current transfer ratio*. It seems odd calling α_0 the common-base current gain when we know *there is no gain*; α_0 is less than unity! We know *why* α_0 is less than unity. It's due to the base current flowing since

$$I_E = I_C + I_B \tag{8.12}$$

and if $I_B > 0$, then

$$I_E > I_C \tag{8.13}$$

which in turn implies

$$\frac{I_C}{I_E} < 1 \tag{8.14}$$

We also know some components that form the total base current:

(a) charge injection from the base into the emitter,
(b) charge recombination in the base.

Component (a) leads to a parameter called the *emitter efficiency*, γ. Component (b) leads to a second parameter called the *base transport factor*, α_T.

Refer again to Figure 8.10. The *emitter efficiency* is a measure of the *injected* electron current compared to the *total* emitter current. The emitter current is given by

$$I_E = I_{En} + I_{Ep} \tag{8.15}$$

Remember, the *current* I_{En} is in the opposite direction to the *electron flow*. The emitter efficiency is therefore given by

$$\gamma = \frac{I_{En}}{I_{En} + I_{Ep}} = \frac{I_{En}}{I_E} \text{ EMITTER EFFICIENCY} \tag{8.16}$$

We can rearrange equation (8.16) to get

$$\gamma = \frac{1}{1 + \dfrac{I_{Ep}}{I_{En}}} \tag{8.17}$$

Since we're dealing with an n^+ pn transistor, $I_{En} \gg I_{Ep}$ since n(emitter) $\gg$ p(base). This gives

$$\gamma \approx 1 - \frac{I_{Ep}}{I_{En}} \tag{8.18}$$

by binomially expanding (8.17) to the first term.

We'll continue by deriving an equation for γ that involves carrier concentration, diffusion coefficient and diffusion length. We're only looking at the emitter–base junction for the moment, so we have all the p–n junction equations of Chapter 6 at our disposal. We know what the injected hole current density into the emitter is; it's just

$$J_p = \frac{qD_p p_{n0}}{L_p} \left(\exp\left[\frac{qV}{kT}\right] - 1 \right) \tag{6.54}$$

or, in terms of current,

$$I_p = \frac{qD_p p_{n0}A}{L_{Ep}} \left(\exp\left[\frac{qV}{kT}\right] - 1 \right) = I_{Ep} \tag{8.19}$$

where

L_{Ep} is the diffusion length of holes *in* the emitter,
A is the emitter–base junction cross-sectional area.

The injected electron current from the emitter into the base comes directly from equation (8.1). Multiplying by the junction area A gives

$$I_n = -\frac{qAD_n n_{p0}}{W_B} \left(\exp\left[\frac{qV}{kT}\right] - 1 \right) = I_{En} \tag{8.20}$$

where W_B is the *effective base width* of the transistor. We'll see why the effective base width and the actual base width of a transistor may differ in the next section.

Using our definition of γ given in (8.18) we can now substitute the results of equations (8.19) and (8.20). This gives

$$\gamma = 1 - \frac{D_p p_{n0} W_B}{D_n n_{p0} L_{Ep}} \text{ npn transistor} \tag{8.21}$$

Using equation (6.37) we can rewrite equation (8.21) as

$$\gamma = 1 - \frac{D_p N_A (BASE) W_B}{D_n N_D (EMITTER) L_{Ep}} \quad \text{npn transistor} \quad (8.22)$$

You should be able to write down the corresponding equation for a pnp transistor immediately.

We now have a mathematical formulation of what we already knew qualitatively, namely that for a good emitter efficiency we need the following:

(a) a high emitter doping (low base doping),
(b) a small base width.

Having exhausted the emitter efficiency, we'll now look at the second parameter I mentioned, the *base transport factor*, α_T.

The *base transport factor* (for an npn transistor) is the ratio of the electron current reaching the collector (I_{Cn}), to the electron current injected from the emitter (I_{En}):

$$I_{Cn} = \alpha_T I_{En} \quad (8.23)$$

This factor gives us some idea of the fraction of injected carriers we lose (by recombination) in sending them across the base. For the following derivation we will be using Figure 8.11. This diagram is not new to you. It's basically the same as Figure 8.6, which we used for a calculation. The main thing I want to get from Figure 8.11 is the *amount* of *excess charge in the base*. Notice the linear drop off in injected carrier density with distance into the base, implying a very thin base (much shorter than the electron diffusion length). From the diagram the excess injected electron charge in the base is going to be

$$q_{BASE} = \frac{q \delta_n A W_B}{2} \quad (8.24)$$

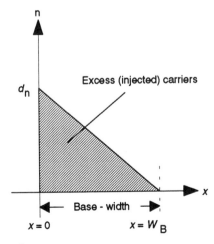

Figure 8.11   Excess base charge.

where

q_{BASE} is the excess stored charge in the base (C),
δ_n is the excess carrier density injected at the emitter–base junction $(x = 0)$,
A is the transistor cross-sectional area,
W_B is the effective base width of the transistor.

What we need now are a couple of expressions for I_{En} and I_{Cn}. We can use the work we've just covered on emitter efficiency for this. From equation (8.16):

$$I_{En} = \gamma I_E \tag{8.25}$$

which gives us the electron current leaving the emitter. We now want the electron current entering the collector I_{Cn}. Combining equations (8.23) and (8.25) we get

$$I_{Cn} = \alpha_T I_{En} = \alpha_T \gamma I_E \tag{8.26}$$

The difference between equations (8.25) and (8.26) is the base recombination current $I_{B(RECOMB)}$:

$$I_{En} - I_{Cn} = I_{B(RECOMB)} = (1 - \alpha_T)\gamma I_E \tag{8.27}$$

You can see looking at equation (8.27) that if we have expressions for $I_{B(RECOMB)}$ and γI_E, then we can get an expression for the base transport factor. I will quickly show you the form of the base transport factor in terms of two *characteristic times* and then I'll convert these times into the transistor base width, and the minority carrier diffusion length. It will become clearer with the maths for a change!

From equation (8.8) we know that you get a current if you look at the rate of change of excess charge with time. We can therefore look at the base and emitter currents in terms of the rate of change of the excess base charge with time. The relevant equations are

$$I_E = \frac{q_{BASE}}{\tau_{TR}} \tag{8.28}$$

and

$$I_B = \frac{q_{BASE}}{\tau_B} \tag{8.29}$$

where

τ_{TR} is the transit time for electrons to cross the base (s),
τ_B is the minority carrier lifetime in the base (s).

Equation (8.29) should be more obvious than (8.28). The base current will depend upon the rate at which minority carriers recombine, which is directly related to the minority carrier lifetime. It's not so clear why the base transit time should be the characteristic time for the emitter current, but I hope to see why it is reasonable a little later. If we now substitute these equations for I_E and I_B into equation (8.27) using the valid assumption that γ is close to unity, then

$$\alpha_T = 1 - \frac{\tau_{TR}}{\tau_B} \quad \text{BASE TRANSPORT FACTOR} \tag{8.30}$$

That's the interim expression. We now want the one with just the base width and the electron diffusion length.

We first need an expression for the emitter current. If we go back to equation (4.44), we see that for the present situation we need to modify (4.44) to get

$$I_n = qD_nA\frac{d\delta n}{dx}\bigg|_{x=0} = I_{En} \tag{8.31}$$

This gives us the current injected at $x = 0$ by the emitter. In other words this is I_{En}. Since we are considering a linear decrease of the injected electron current the differential evaluated at $x = 0$ is simply

$$\frac{d\delta n}{dx}\bigg|_{x=0} = \delta n/W_B \tag{8.32}$$

Refer back to Figure 8.11 to check that this is OK. Lastly, since $\gamma I_E = I_{En}$ and gamma is close to unity, we can combine equations (8.31) and (8.32) to give

$$I_E = \frac{qD_n\delta_n A}{W_B} \tag{8.33}$$

We can get at the emitter current another way. The emitter current multiplied by the time it takes the injected electrons to cross the base must equal the excess charge in the base. The excess charge in the base is given by (8.24), so this second way of getting at the emitter current gives

$$\tau_{TR}I_E = \frac{q\delta n W_B A}{2} \tag{8.34}$$

Combining (8.33) and (8.34) we can get an equation for τ_{TR}:

$$\tau_{TR} = \frac{W_B^2}{2D_n} \tag{8.35}$$

The diffusion length is related to the minority carrier lifetime as given by the expressions following equation (4.67). So for the minority carrier lifetime in the base we can write

$$\tau_B = \frac{L_{nB}^2}{2D_n} \tag{8.36}$$

where L_{nB} is the electron (minority carrier) diffusion length in the base. The final step is to substitute (8.35) and (8.36) into (8.30), which yields

$$\alpha_T = 1 - \frac{W_B^2}{2L_{nB}^2} \quad \text{BASE TRANSPORT FACTOR} \tag{8.37}$$

After all that was it really worth it? Equation (8.37) tells us that if we don't want much recombination in the base, then

(a) make the base narrow (small W_B),
(b) ensure a long minority carrier diffusion length in the base (large L_{nB}).

Qualitatively, we already knew this, and it took a lot of maths to get this result. Sometimes a lot of involved maths throws up exactly the same answer as a little thought with the right picture!

The final derivation I want to cover in this section is the relationship between the characteristic times we discussed above, and the *common-emitter current gain* β. This derivation should start making you think about gain in the photoconductor. Why? Recall the gain mechanism in the photoconductor. If the carrier lifetime in the photoconductor was greater than the transit time across the photoconductor, we could get gain. You'll see that we come up with an equation of *exactly the same form* for the bipolar transistor if the gain is governed by recombination in the base, rather than by hole injection into the emitter.

If the current gain of a bipolar transistor in the common-emitter configuration is controlled mostly by *recombination in the base*, then the following derivation is valid. The common-emitter current gain β is the collector current divided by the base current. Since the collector current is close to the emitter current for a good transistor, we can use equation (8.28) to get

$$I_C = \frac{q_{BASE}}{\tau_{TR}} \qquad (8.38)$$

Combine (8.38) with (8.29) to get

$$\beta = \frac{I_C}{I_B} = \frac{\tau_B}{\tau_{TR}} \quad \text{COMMON-EMITTER CURRENT GAIN} \qquad (8.39)$$

Exactly the same result as for the photoconductor!! I must stress that this is only the case if recombination in the base is dominant. If hole injection into the emitter is the major component of the base current, then we did a calculation near the beginning of this chapter to see what the common-emitter current gain was under these conditions. Recall that the gain was close to the emitter doping divided by the base doping (quite a big number, usually in the thousands). If we now consider the gains we get from equation (8.39) we're more likely to get gains in the hundreds instead.

If, under the above conditions, our transistor really is behaving like the photoconductor, you should be thinking about speed/gain trade-offs just as we did with the photoconductor. One speed limitation with our transistor is the charge stored in the base (diffusion capacitance). Look at equation (8.9). That's OK. We can drastically *cut down* the diffusion capacitance by putting a high concentration of *lifetime killers* in the base! That's speeded up our transistor. Now look at equation (8.39). Killing the base lifetime kills the current gain! It *really* is just like the photoconductor.

My last comment on the base recombination current is this. For the thin bases we normally use in integrated circuits, base recombination current really doesn't bother us. Recombination current *will* bother us (look at equation (8.37)) if the base is wide, and if the base has been lifetime killed (this will drop the minority carrier diffusion length via equation (8.36)). These conditions are precisely what we find in *power bipolar transistors*, so in these devices the base recombination current *is* likely to dominate.

Calculation

Derive a simple expression relating the common-base current gain, α_0, to the common-emitter current gain β.

Considering small signals (small current changes) we can write

$$\beta = \frac{I_C}{I_B}, \quad \alpha_0 = \frac{I_C}{I_E}$$

and from equation (8.12):

$$I_E = I_C + I_B$$

The common-base current gain can therefore be rewritten as

$$\alpha_0 = \frac{I_C}{I_E} = \frac{I_C}{I_B + I_C} = \frac{1}{(I_B/I_C) + 1} = \frac{\beta}{1 + \beta}$$

A similar analysis of the common-emitter current gain yields

$$\beta = \frac{\alpha_0}{1 - \alpha_0}$$

8.6 Punch-through

In the active mode we operate the collector–base junction under reverse bias. This means the depletion-layer width is wider than in the thermal equilibrium (no bias) case, and it will get wider with increasing reverse bias. Figure 8.12 shows the situation. In Figure 8.12(a) the npn transistor is operating in the active mode, with a reverse bias of V_1 across the collector–base junction giving a depletion-layer width of W_2. The depletion-layer width of the forward biased emitter–base junction is W_1. If the collector–base reverse bias is increased to V_2, as shown in Figure 8.12(b), then, clearly, the collector–base depletion-layer width will widen, *and as a consequence the neutral base width will be reduced*. This *neutral base width* is what I've been calling the *effective base width*. It's the region over which electron *diffusion* occurs. But if the effective base width W_B is changing with V_{CB}, then equation (8.20) tells us that the emitter current I_{En} will change as well! As V_{CB} increases, W_B decreases and I_{En} will increase. This allows us to understand the *I–V* characteristics of the transistor in this mode which is shown in Figure 8.13. The *I–V* characteristic is much as we would expect for the forward biased emitter–base p–n junction. However, note how the emitter current increases with increasing V_{CB} due to the *decreasing* effective base width. We're clearly going to hit a problem if we keep increasing V_{CB}. The depletion-layer width will increase as $V_{CB}^{0.5}$ (look at equation (6.61)), and for *narrow base widths* (high-speed transistors) it will eventually reach the edge of the *emitter–base* depletion layer. At this point we have *punch-through*. The *emitter* current will rise directly with V_{CB}, and normal transistor action ceases. For wider base widths it is likely that breakdown of the collector–base junction (see Section 6.9) will occur before we reach the punch-through condition.

8.7 Modes of operation

The bipolar junction transistor has *two* junctions which may either be *forward* or *reverse* biased. This leads to four possible bias configurations, or, *modes of operation*. The four operating modes are as follows:

(a) active,
(b) saturation,
(c) cut-off,
(d) inverted.

The *active mode* is the mode of operation we've considered so far. Figure 8.14 shows the bias configuration, and minority carrier distributions, for the transistor operating in this mode. The emitter–base junction is forward biased, and the collector–base junction is reverse biased.

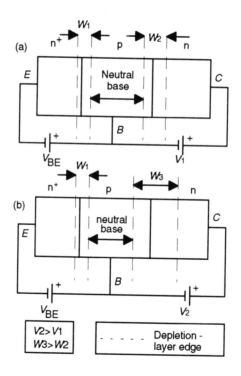

Figure 8.12   Neutral base-width.

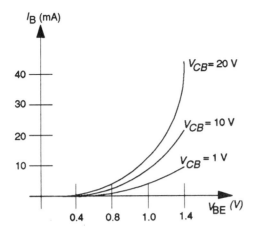

Figure 8.13   npn *I–V* characteristic common-base connection.

In the *saturation mode* both junctions are forward biased, as shown in Figure 8.15. In this mode the transistor is in a highly conducting state and behaves as though it were a closed switch. In the saturation mode V_{CE} is small, virtually no volts are dropped across the transistor, hence the transistor is *on*.

The *cut-off* mode (Figure 8.16) is relatively straightforward to understand. Both junctions are reverse biased and the only currents flowing are the *very small* reverse leakage currents. The cut-off mode is therefore the converse of the saturation mode. The transistor acts like an open (off) switch.

The *inverted mode* is also called the *inverted active mode* because, in this instance, the emitter–base junction is *reverse biased*, and the collector–base junction is *forward biased*. The configuration is shown in Figure 8.17. As we have an emitter acting like a collector, and a collector acting like an emitter, it is not surprising that we are going to get poor current gain with this mode. Why? Because the 'emitter efficiency' is going to be very low since the layer dopings are the 'wrong way round' (the 'emitter' is low-doped and the 'collector' is high-doped in the inverted mode). Although this may seem a strange mode of operation it *is* utilized in I^2L (integrated injection logic) circuits.

Figure 8.18 shows the *I–V* characteristic of an npn transistor connected in the common-emitter configuration. This is the configuration that is used most frequently as we get large changes in collector current for small changes in the base current (current amplification). On the *I–V* characteristic I have shown three of the four operating mode regions. Convince yourself that the saturation region is indicated correctly (you want small V_{CE}, large I_C). When you first see this characteristic you intuitively feel that the saturation region is where the collector current curves go horizontal. *It isn't.*

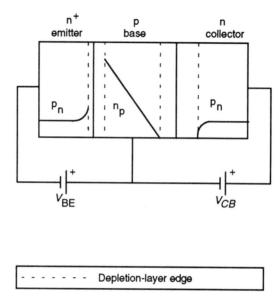

Figure 8.14   npn transistor in the active mode.

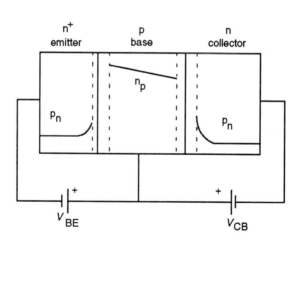

Figure 8.15   npn transistor in the saturation mode.

8.8 Two simple circuits

I will very quickly show you two simple circuits using a single npn transistor, which you can build. The first circuit is used to illustrate what I was trying to do way back in Figure 1.1. The second circuit is to show you how you *should* bias the transistor properly!

Look at Figure 8.19. The transistor is any small-signal npn bipolar transistor with reasonable gain, such as ZTX300, ZTX450, ZTX605, etc., etc. The red LED is any low-current type that operates with about 10 mA forward current. The switch can connect the base resistor to either the positive rail (+9 V), or earth (0 V). If we want 9 mA to flow through the LED we need a current limiting resistor of 1 K as shown, and we need the transistor to be biased *on*. The base current needed to supply 9 mA collector current can be calculated from the transistor's β. Usually, the β given by the manufacturers covers quite a large range. If you pick the *smallest* current gain (in the range given) and use

base current required = collector current divided by smallest β

you should get a suitable working value for the base current. Please note, *this is not the proper way to do this!!!* I'm only showing this first circuit for illustration. The current gain β is *not a good parameter* (as stated, it can have a very wide range of values) so you shouldn't *rely* on knowing β when designing a real circuit. Let's carry on regardless. I'm *assuming* (bad idea) a minimum current gain of 50 so I'll need 180 μA flowing into the base which I can provide with a base resistor of 50 kΩ ($50 \times 10^3 \Omega$) going to the +9 V rail. With the switch in the *up* position the LED will shine brightly, and with the switch in the down position (or disconnected) the LED will go out. Clearly, with the base disconnected the base cannot be forward biased, and so no transistor action occurs. The same reasoning applies to the situation where the base is connected to the ground (0 V) rail (the base is then shorted to the emitter). This is what I was trying to do all those years ago. You can see in Figure 8.1 that without the current limiting resistors in the base and collector, the only current limiting element I had was the torch bulb. Hence with the transistor *on* too much current flowed, and the transistor 'burned out'.

Now look at Figure 8.20 to see how you *should* bias the base properly using a resistive divider. This arrangement makes the circuit less sensitive to changes in β that can occur with changes in collector voltage V_{CC}. I've also changed the collector and emitter circuits a bit. By moving the *load* (LED plus resistor) down into the emitter arm, and connecting the collector directly to the positive rail, I've formed what's called an *emitter-follower* circuit. In this configuration the *emitter* voltage closely *follows* the voltage on the *base* (the emitter voltage is in fact the base voltage *minus* the base–emitter voltage drop). So, suppose we choose an emitter voltage of 3 volts. To get 9 mA flowing in the emitter arm we need a resistor of 300 Ω (I'm assuming the LED resistance is zero, but *you* can take a

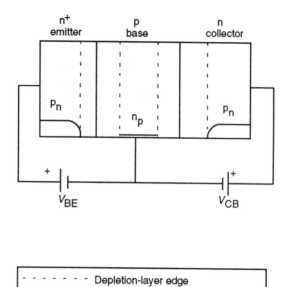

Figure 8.16   npn transistor in the cut-off mode.

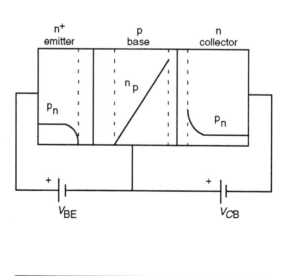

Figure 8.17   npn transistor in the inverted mode.

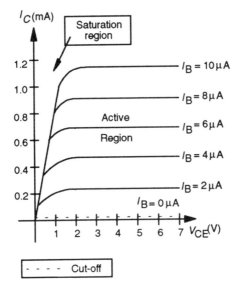

Figure 8.18 npn *I–V* characteristic common emitter connection.

non-zero resistance into account when *you design* your own circuit!). If we
consider a junction voltage drop of 0.6 V (no, it's not always valid physically, but
it usually tends to work well in circuit design) then for 3 volts at the emitter we
need 3.6 volts on the base. Now *lots* of divider network resistor values will give us
3.6 volts on the base, but we'll use the criterion that the divider input impedance
must be much less than β times the value of the emitter resistor. Unfortunately, to
see why this is so you must go to a circuit book, such as the excellent text referred

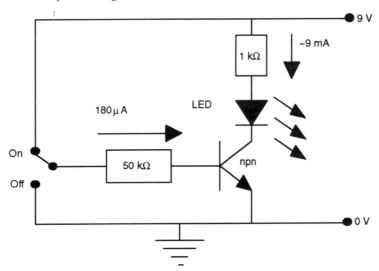

Figure 8.19 LED on-off circuit.

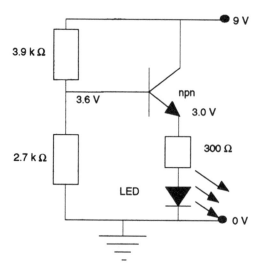

Figure 8.20 Correct base biasing.

to in Further Reading. If we choose a (minimum) β of 50 again, then our resistive divider input impedance must look less than $15\,k\Omega$. I'll choose a standard resistor value for the lower arm of the divider of $2.7\,k\Omega$ which will make the upper resistor value about $4.1\,k\Omega$ (to get $3.6\,V$ on the base). The nearest (lower) standard resistor value is $3.9\,k\Omega$, so I'll put this in the upper arm. Now the (parallel) input impedance of the divider chain is about $2\,k\Omega$, which satisfies our impedance criterion. The base voltage will be $3.68\,$volts giving $3.08\,$volts at the emitter, which is fine. Connect up the battery and the LED *will* light!

8.9 HJBT and polyemitter

In today's integrated circuit bipolar transistors there is very little carrier recombination in the transistor base. Virtually all the loss in current gain is from carrier injection from the base into the emitter. As this seems to be such an important gain reduction mechanism it's worth looking at ways of suppressing minority carrier injection into the emitter. Two ways of doing this utilize *heterojunctions* and *polysilicon emitters* (polyemitters).

Figure 8.21 shows the energy-band diagrams for

(a) an ordinary (homojunction) npn bipolar transistor,
(b) an npn bipolar transistor with a wide band-gap emitter,
(c) an npn transistor with a narrow band-gap base.

Transistors (b) and (c) are *heterojunction* bipolar transistors (HJBTs) as they have *different materials* on either side of the emitter–base junction.

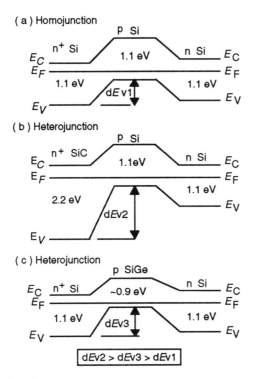

Figure 8.21   Heterojunctions.

Look at the emitter–base junction in each case. For the standard bipolar transistor (a) nothing prevents the injection of holes from the base into the emitter in the normal (forward biased p–n junction) way. Now look at (b). The wide band-gap emitter (which could be formed in silicon carbide) presents *an increased hole barrier height* to hole injection from the base into the emitter. Since the number density of injected holes will *decrease exponentially* with *linear increase* in barrier height, it only requires a *small* barrier height increase to *drastically reduce* the number of injected holes. This in turn will give us an enormous increase in current gain! A similar increase in valence band barrier height is experienced if we use a narrow band-gap base material as in (c). In this case the narrow band-gap base could be made in the alloy *silicon–germanium* (SiGe), with exactly the same result. In both cases we get reduced hole injection from base to emitter, and therefore increased current gain. A major difference in structure between the two approaches is that the wide band-gap emitter only introduces *one* heterojunction, where it is needed, at the emitter–base junction. The narrow band-gap SiGe approach introduces *two heterojunctions*, and the second (unnecessary) heterojunction at the collector–base junction can cause problems if we're not careful. We can now trade-off this increase in gain against other transistor parameters if we want. For instance, we can now *increase* the base doping, and bring the gain back to what it was with the ordinary homojunction bipolar transistor. Why should we want to do that? Because the increased base doping will result in a *decrease* in the base resistance, which in turn will produce a *faster* device. We can thus trade-off *speed* against *gain* in the HJBT. I'd better quickly spell out what this base resistance is because it's not intuitively obvious. The base resistance I'm referring to *is not* the resistance to electrons flowing *across* the base. It's far more fundamental than that. Figure 8.22 shows a schematic planar bipolar transistor structure. For a planar structure you've somehow got to make contact with the base region *from* the top surface of the silicon. This entails bringing connections out sideways from the base region that can link up with metal contacts on the surface. *It is the resistance introduced trying to make contact to the base region in this way that leads to the base resistance.* So the base resistance is a *lateral* base resistance arising from this (technological) problem of contacting the base layer.

Another way of suppressing hole injection into the emitter is with the polyemitter structure shown in Figure 8.23. In this structure we have a very thin oxide (SiO_2) layer (of order 1 nm) separating the base and emitter regions. Because we are growing the silicon emitter on oxide (rather than single-crystal silicon), which is amorphous, there is no underlying crystal structure for the emitter growth to follow, and it grows as polycrystalline silicon (polysilicon) when we're making the transistor. The thin oxide layer has a very interesting property, it allows electrons to pass through it more easily than holes! Again the result is hole suppression, and a large increase in current gain. Just a couple of points about the oxide before we move on. The oxide has a big band gap (several eV) and it is *very* thin, so carriers get through this oxide layer by quantum mechanical tunnelling. The precise mechanism of tunnelling is too involved for us to study here, but the fact is although we've put a dielectric (insulator) in the way, it doesn't upset the electron current flow too much (it doesn't add a lot of *emitter resistance*). Although it doesn't upset the electron flow, it *does* reduce the hole current considerably, because the tunnelling coefficient is much higher for electrons than for holes. The basic mechanism for gain enhancement in polysilicon emitter structures is this difference in tunnelling probability, for electrons and holes, through a thin oxide layer.

8.10 Vacuum microelectronics

I hope by now you appreciate that the transistor is a pretty neat device. However, for the job they've got to do, transistors have a lot of undesirable properties.

We've seen that transistors are bipolar devices. They function by moving around *both* electrons and holes. What's pretty bad is that they're trying to move these carriers around in a semiconductor crystal lattice which gets in their way. We know that if you try to move charge carriers around inside semiconductors

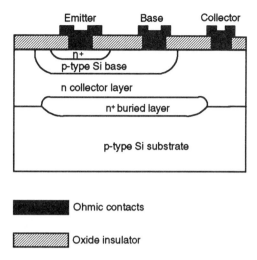

Figure 8.22 npn planar bipolar transistor.

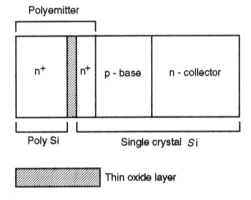

Figure 8.23 Polyemitter structure.

using an electric field, you come up against a limiting velocity called the saturation drift velocity. This is not very useful when trying to make high-speed devices. You can use materials other than silicon to get higher drift velocities, but the velocities are still pretty low compared to the ultimate velocity: the velocity of light *in-vacuo*. There's worse to come. The fact that the velocities of the charge carriers are limited by collisions with the lattice means that the crystal lattice heats up. This gives us the additional problem of device destruction due to overheating if the currents get too large. It doesn't end there. Since *both* electrons and holes are involved you get unwanted minority carrier effects (diffusion capacitance) which also slow the device down. This is all adding up to trying to make the best out of a bad job, especially since there's a much more promising electronic device that doesn't have *any* of the above problems.

There *is* an electronic device in which the charge carriers, electrons in this case, can travel at near-light velocities. The electrons are not hindered in their movements by an unnecessary crystal lattice. Consequently there is no crystal lattice to damage by Ohmic heating. With no obstructing crystal lattice there is no material present to show adverse temperature effects, such as going *intrinsic* at moderately low temperatures. For high-radiation environments, such as space or nuclear reactors, the absence of a crystal lattice means no radiation damage and no soft errors in memory devices. What is this amazing electronic device? It used to be called a bottle, or a tube, or, most commonly, *a valve*.

Has the struggle of trying to get this book together been too much? Has Parker flipped? I hope to show you that reconsidering the valve is not a pointless exercise at all. Historical analysis is vitally important in our work. You can see how things have evolved to their present state of development, and you can see that the evolutionary route taken was not necessarily the best one. Talking about Evolution, which came first, the crystal rectifier or the valve? In fact it was the *crystal*, but the way the initial crystal rectifiers were manufactured meant that the 'new valve' was a lot more reliable and effective. Valves took over. We then had to wait a while for Bardeen, Brattain and Shockley to come along and go back to the work on crystals. The transistor was invented, and it had many useful properties that would displace the valve for virtually all electronic applications. Advantages of the transistor were that it was smaller, and later versions were more robust than the valve (I don't think the *original* point contact transistor with great wires hanging about all over the place would have been much more robust than the valve). The power consumption of the transistor was *much less* than the valve. One reason for this was that the transistor did not need a power-consuming heater to drive a thermionic (electron-producing) emission process.

The *real killer* for the valve was the way in which transistors could be integrated to provide very large complex circuits in very small, highly reliable packages. This being the historical fact, why even bother thinking about valves again?

Valves at the time of their demise could certainly work at 10 GHz or more, with power handling capabilities of 10 watts or more. Look at present-day semiconductor devices. We're only just beginning to approach this sort of performance! In the case of *very high power* handling, such as TV or radio transmitter stations, valves still have a place, although they are slowly giving way to solid-state devices. For the case of large-area display devices (cathode ray tubes, colour television tubes) valves seem to be holding their own for the moment. Let's go back to the 10 GHz 10 watt valve. Obviously, things were reaching a limit here as it was nearing the end of the valve's dominance. What was the problem? To make valves *fast* is easy: you make them *smaller*. But the manufacturers at that time were reaching the technological limits of valve fabrication. The valve elements were having to be aligned to tolerances of tens of

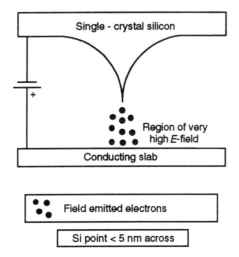

Figure 8.24 Field emission.

microns, which as you can imagine is a little difficult with metal foils and glass envelopes. So why should things be any easier now? It's because the transistor has provided us with a *fabrication technology* that can be used to make very small (micron-size) structures. But what about the power dissipation problem? The use of thermionic emission to produce electrons in *very* small devices wouldn't be a good idea. Is there another electron emission mechanism we can use? Fortunately there is; we can use *field-emission*. The intense electric field that can be produced in the vicinity of a sharp conducting point can literally pull electrons out of the material. Figure 8.24 shows the field-emission of electrons from a very sharp point formed in silicon. The sharp point is fabricated using advanced etching techniques.

It looks as if we have all the basic elements. We have a low-power (field-emission) electron source, and the means to make very small devices. Where are the microvalves? Strange as it may seem this is a relatively new development, so the microvalves are only just beginning to appear. The subject area itself is called *vacuum microelectronics* and researchers are developing some very clever (micromachining) techniques to form these very small devices. The cross-section of a *microelectronic valve* (of order 100 μm high) is shown in Figure 8.25. To form this sort of structure requires very precise growth, and etching, of silicon and silicon dioxide.

You must be aware that technology often seems to go 'full circle'. Having seen this demonstrated by these new microelectronic valves, it may be worth while looking at other 'old' or discarded technologies, to see if they can be resurrected by today's technological advancements.

BIPOLAR TRANSISTOR

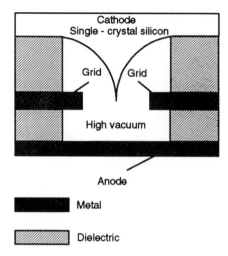

Figure 8.25   Microelectronic 'valve'.

CHAPTER NINE

Field-effect transistors

The shock of coming across a *three*-terminal device should have worn off by the end of the last chapter. Here is another three-terminal device; in fact it's another transistor. The field-effect transistor we shall concentrate on is the metal oxide semiconductor (MOS) transistor, the transistor that is usually used in very large-scale integrated circuits (VLSIs), microprocessors and the like. There are a few significant differences between the bipolar junction transistor (BJT) of the last chapter and the MOS transistor, even though the *I–V* output characteristics of both look very similar. Recall that the BJT was a *current* controlled device. The voltages were only there to drive the currents. The MOS transistor, however, is a *voltage controlled* device. By this I mean the device switches, or amplifies, due to the application of a voltage source and (practically) no current is required. Another major difference between the two devices is that the BJT is a bipolar device requiring both electrons and holes to operate. The MOS transistor, however, is *unipolar*, transporting only one carrier type. A third difference between BJTs and MOSFETs is that current flow in the MOSFET is *drift* dominated, whereas current flow in the BJT is a *diffusion* dominated process.

Again, this chapter will make a break from the 'standard' text formats. Quite often the 'standard' texts discuss the junction field-effect transistor (JFET) first, and then move on to the metal-oxide semiconductor field-effect transistor (MOSFET) afterwards. I have not done this for a couple of reasons. One is that although many of the equations are *similar* for both devices, the actual operation of the two types of device is quite different. I think that 'linking' both devices *closely* together in the same chapter could cause confusion. This leads to the obvious solution of putting each device in its own chapter. I don't want to do that because:

(a) it would lengthen the book beyond my own preconceived limit, and
(b) I don't believe the JFET is of sufficient current technological importance to warrant its own chapter. My apologies to the designers and manufacturers of JFETs, but it is the MOSFET that is at the heart of the very large-scale integrated circuits (VLSIs) of today. Having said all that I believe the JFET

has sufficient importance that it cannot be omitted altogether, hence the JFET makes a brief appearance towards the end of this chapter.

Because there are so many different varieties of FET I have produced Figure 9.1, which lists the main types (I will not cover the MESFET in this introductory text). However, you should note that you may not come across the discrete p-channel variant (where *discrete* means the packaged device, on its own, *not* in an integrated circuit with other transistors). Why is this? Recall that these FETs are unipolar devices and carrier movement is *drift dominated*. This being the case we will get the *fastest* devices if we move electrons around rather than holes (just compare the *drift mobilities* of electrons and holes in Si *and* GaAs). You *will* come across both n- and p-channel variants in integrated circuits (ICs) because there are many circuit advantages to be gained in using the complementary (n- and p-channel) devices together in the same circuit. Such circuits are called CMOS (complementary metal oxide semiconductor) circuits when using the metal oxide semiconductor system.

In *exactly* the same way that our study of the BJT was made easier with foreknowledge of the p–n junction, we'll find our understanding of the MOSFET is made simpler by first studying the MOS diode structure.

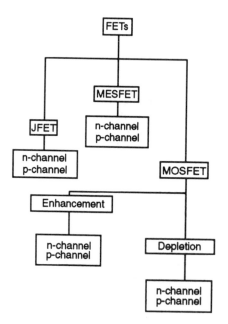

Figure 9.1 Types of FET.

9.1 The MOS diode in thermal equilibrium

Figure 9.2 shows the MOS diode structure, and the energy-band diagram for the device in thermal equilibrium. As with the Gunn diode, this device is *not* a diode in the sense of being a rectifying element. *Diode* again simply means a two-terminal device. Far from being a rectifying element, the MOS diode is in fact a funny sort of capacitor as we'll see as this chapter progresses. For this reason the *MOS diode* is also known as the *MOS capacitor*. Return to Figure 9.2, where a couple of things need some explanation. The dotted lines in the oxide energy bands are there to imply that a section has been removed so that we can see the flat top of the oxide energy band on the page. The band-gap energy of SiO_2 (about 9 eV) is so much bigger than silicon (1.1 eV) that if Figure 9.2 was drawn to scale, the top of the oxide band would go off the top of the page. Notice that the top of the oxide band is flat and horizontal. We know what this means; it means there is no voltage drop (electric field) across the oxide. When we start applying some volts to the metal (we will assume that the semiconductor is held at earth potential) we shall see that the voltage drop is taken up by a slope in the energy band of the oxide layer. Since I have short-circuited the metal to the semiconductor, the Fermi-levels *must* be aligned (at the same height), as there is no tendency for current flow in either direction. Finally, note that the bands in the semiconductor under these thermal equilibrium conditions are flat and horizontal. Perhaps not too surprisingly, this is known as the *flat-band* condition. Physically, it means that there are no electric fields (or charge distributions) which will bend the bands in this thermal equilibrium case. In reality *there may be* charges in the oxide (unintentionally introduced) which *will* cause band bending in the semiconductor, even under thermal equilibrium conditions. There are also active electronic states at the semiconductor–oxide interface (interface states) but I don't want to go too deep into the physics of the semiconductor–oxide interface. This is why Figure 9.2 is labelled an *ideal* MOS diode. It doesn't consider oxide charges or interface states or 'leaky' (slightly conducting) oxides. A final point before leaving discussion of Figure 9.2. Suppose we *do* have some charge in the oxide which is causing band bending in the semiconductor. We could then apply some volts to the metal (to counteract the effects of this oxide charge) and the bands in the semiconductor would flatten out again. The voltage we have applied is called the *flat-band voltage* and we can measure this voltage for MOS diodes to get some idea of how much charge there is in the oxide (which gives us an idea of the oxide quality). I know you won't *fully* appreciate what's just been said yet, but it *will* become clearer once we start applying volts to the metal contact of the MOS diode. Let's get on with it.

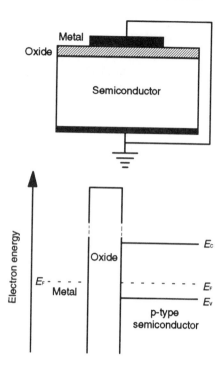

Figure 9.2   Ideal MOS diode in thermal equilibrium.

9.2   The MOS diode with applied bias

Figure 9.3 shows the MOS diode structure for an n–Si substrate with a bias voltage applied to the metal top contact. The three conditions we will be considering are *depletion, inversion and accumulation*. I will go through a discussion of these conditions using the simple MOS structure *before* looking at the relevant band diagrams and equations.

Consider the depletion condition in Figure 9.3 first. We have applied a *small* negative potential to the top metal contact which will put a small negative charge on this contact. The oxide layer prevents current flow but the effect of the negative charge is felt by the electrons in the n-type silicon. The negative charge tends to repel the free electrons from a region underneath the top contact, as shown, forming a *depletion region*. This is exactly the same sort of depletion region as we saw in the p–n junction. It is a region *depleted* of free charge carriers. Look at the MOS structure again. The top metal contact and the oxide layer together look like half of a capacitor structure. The semiconductor with its free charge carriers forms the 'other plate' of the capacitor. That's why the device is also called an MOS capacitor. Now look what we've done by putting the MOS diode into depletion. We've pushed the charge carriers away from the interface and effectively we've *increased the separation* of the capacitor 'plates'. Why? Because the depleted semiconductor region (no free charges) acts like a dielectric and so it looks like we've widened the oxide layer (although the depletion region *will* have a different *permittivity* from the oxide layer). Consequently, the capacitance of the MOS diode will *decrease* as we push the diode into depletion. It decreases *from* a value we can calculate. The oxide capacitance of the MOS diode is just the parallel plate capacitance for a two-plate capacitor separated by an oxide dielectric:

$$C_{OX} = \varepsilon_{OX}\frac{A}{d_{OX}} \qquad (9.1)$$

where

C_{OX} is the oxide capacitance (Farad),
A is the capacitor area (cm^2),
d_{OX} is thickness of the oxide layer (cm),

and

$$\varepsilon_{OX} = \varepsilon_0 \cdot \varepsilon_{r(OX)} \qquad (9.2)$$

where

ε_{OX} is the oxide permittivity (F/cm),
ε_0 is the permittivity of free space (8.854×10^{-14} F/cm),
$\varepsilon_{r(OX)}$ is the relative permittivity of oxide (3.8).

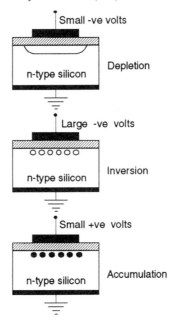

Figure 9.3   Ideal MOS diode.

Calculation

Calculate the oxide capacitance of an MOS diode if the top metal contact is 2 mm in diameter and the oxide thickness is 100 nm. This is the sort of structure we often use when measuring the flat-band voltage in order to ascertain oxide quality.

Simply substitute the above values into equation (9.1) to get

$$C_{OX} = \frac{8.854 \times 10^{-14} \times 3.9 \times \pi \times (0.10)^2}{10^{-5}} \text{ Farad}$$

$$= 1.1 \text{ nF}$$

If the dielectric material is depleted silicon rather than silicon dioxide, simply replace the oxide permittivity in equation (9.1) by the permittivity of silicon.

Experimentally, we find that we cannot keep increasing the negative potential on the top contact (Figure 9.3) and keep pushing the depletion region further and further into the semiconductor. Eventually, it becomes *energetically favourable* for *holes* to collect at the oxide–semiconductor interface rather than for the depletion-layer width to increase any further. This condition is called *inversion*. We've *inverted* the charge carrier type at the oxide–semiconductor interface! We shall see a little later what the conditions are (mathematically) for inversion, but for now notice that we have a lot of free charge carriers back at the oxide–semiconductor interface (even if they are holes rather than electrons). As far as our MOS capacitor is concerned, it looks like the second 'plate' has come back to the interface! The capacitance is back to the oxide capacitance. Where did these holes come from? They are the minority carriers naturally occurring at a finite temperature in a semiconductor, and they will collect in the *potential well* under the top electrode in inversion. If we formed the depletion region *very quickly* (by applying a very quick voltage pulse) we can push the depletion region into the silicon before the holes have got a chance to collect at the interface forming an inversion layer. By this means we can actually push the depletion region *further* into the silicon than inversion normally allows us to, provided we do it quickly enough. This condition is called *deep-depletion*.

It's almost an anticlimax to consider the last condition, *accumulation*. For our n-type semiconductor in Figure 9.3 we simply apply a positive potential to the top electrode. The positive charge on the top electrode will attract the negative majority carriers in the n-type silicon and we will thus *accumulate majority carriers* at the oxide–semiconductor interface. Again, it looks like the second 'plate' of our capacitor is back at the interface, and so we will measure just the oxide capacitance in accumulation. Figure 9.4 shows a possible capacitance–voltage (C–V) characteristic for the type of diode we've just discussed.

Figure 9.5 shows the MOS diode conditions for a p-type semiconductor substrate. This structure will be of more interest to us (than the n-substrate) when

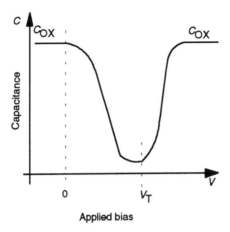

Figure 9.4 Ideal MOS diode $C–V$ characteristic.

we look at the enhancement-mode MOSFET because we actually *use* the *inversion* condition in this device. Remember, we want to move *electrons* around for high speed and if we have electrons present in the *inversion layer* then we must have a *p-type* substrate. This will all become much clearer in the next section.

That was a *qualitative* look at the MOS diode under bias. We'll now look at the appropriate band diagrams, and the relevant equations for the MOS diode, before studying the MOSFET itself.

9.3 MOS diode band diagrams

For the reasons mentioned at the end of the last section, I am going to consider the metal–oxide structure on *p-type* semiconductor substrates only. However, when you see the form of the band diagrams in this case, you'll easily be able to construct your own for n-type substrates. As in the last section, we will go through the depletion, inversion and accumulation conditions in turn.

Let's consider the MOS diode structure, using a p-type substrate, with a small *positive* voltage applied to the top metal electrode. We know that this will lead to a *depletion* of the majority carrier holes at the oxide–semiconductor interface. This is shown in Figure 9.6(a), together with the relevant band diagram in Figure 9.6(b). With an applied positive potential, the Fermi-level in the metal moves downwards from its thermal equilibrium position, as shown. The voltage drop across the oxide is reflected by the slope in the oxide band energy, and the semiconductor bands now bend near the oxide–semiconductor interface. Remembering that holes 'float' and electrons 'sink' under drift conditions, it is clear that with the band bending shown, holes will move away from the interface resulting

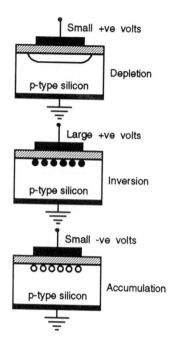

Figure 9.5   Ideal MOS diode.

in a depletion region. The charge conditions are shown in Figure 9.6(c). The uncovered (negative) donor charge (per unit area) is $-qN_AW$ and this just balances the positive charge per unit area Q_M on the metal electrode (assuming no other charge sources are present). The depletion-region width is W. Note that the Fermi-level in the semiconductor is flat and horizontal, as expected, since no current can pass through the oxide layer.

If we now increase the positive potential on the metal electrode to a value above the *threshold voltage* (V_T), the semiconductor surface will *invert* and an (electron) inversion layer will reside at the interface. Figure 9.7 illustrates the MOS diode in the inversion condition. With the increased positive potential the Fermi-level in the metal is depressed even further. The greater voltage drop across the oxide is reflected in the greater slope in the oxide energy band, and there is even greater band bending in the semiconductor. In fact, the band bending in the semiconductor is so great that the intrinsic Fermi-level has *crossed* the Fermi-level. What significance does this have? The electron concentration as a function of $E_F - E_i$ is given by equation (3.9):

$$n = n_i \exp\left(\frac{E_F - E_i}{kT}\right) \tag{3.9}$$

and since

$$E_F - E_i > 0 \tag{9.3}$$

then

$$n > n_i \qquad (9.4)$$

$$p < n_i \qquad (9.5)$$

and the surface of the semiconductor is *inverted* (there are *more electrons* than *holes*, but we're in p-type material!). The thickness of the inversion layer is *very small*, typically 1–10 nm.

I have called the depletion-layer width W_m in this inversion case, W_m standing for *maximum* depletion-layer width. The reason for this is once the inversion layer has formed, the depletion-layer width reaches a maximum. *Any further charge increase appears in the inversion-layer charge only.*

Figure 9.7(c) shows the *positive* top metal electrode charge Q_M being balanced by the *negative* inversion-layer and depletion-layer charges, i.e.

$$Q_M = qN_A W + Q_n \qquad (9.6)$$

The final situation to consider in this section is *accumulation*. For a negative potential applied to the top metal electrode the appropriate diagrams are given in Figure 9.8. The Fermi-level in the metal rises above the thermal equilibrium position. The slope in the oxide band now corresponds to a voltage drop of opposite polarity to that considered in the last two cases. The bands near the

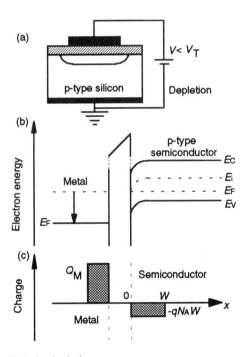

Figure 9.6 Ideal MOS diode in depletion.

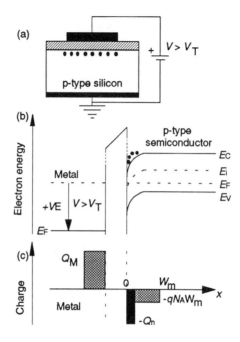

Figure 9.7   Ideal MOS diode in inversion.

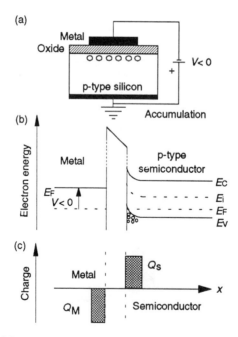

Figure 9.8   Ideal MOS diode in accumulation.

interface, in the semiconductor, now bend *upward*. As in the other cases there is no current flow through the oxide so the Fermi-level in the semiconductor remains constant. Equation (3.12) tells us how the majority carrier hole concentration, p, varies with $E_i - E_F$, namely,

$$p = n_i \exp\left(\frac{E_i - E_F}{kT}\right) \qquad (3.12)$$

and, as upward band bending causes $E_i - E_F$ to *increase*, then the hole density *increases*. The increase of p at the surface with increasing negative potential is of course just the accumulation condition. The *negative* charge on the metal Q_M is balanced by the *positive* hole charge Q_S (Figure 9.8(c)) in this instance.

Derivation

Figure 9.9 shows an MOS diode, formed on a p-type substrate, in the condition known as *strong inversion*. If we define the strong inversion condition as requiring E_i to lie as far below E_F at the *surface* as it does *above* E_F far away from the surface, write down an equation for V_{INV} (strong inversion).

Go back to the p–n junction, Section 6.2, and refamiliarize yourself with that work. In that section we derived equation (6.8), which relates the hole potential V_p to the energy difference between the intrinsic Fermi-level and the Fermi-level. In other words, the V_p of equation (6.8) is the same V_p we're considering in Figure 9.9. This being so, we can now derive the equation for V_{INV}. Starting with equation (6.8):

$$|V_p| = \frac{kT}{q} \ln\frac{N_A}{n_i} \qquad (6.8)$$

and since

$$|V_{INV}| = 2|V_p|$$

then

$$|V_{INV}| = \frac{2kT}{q} \ln\frac{N_A}{n_i} \quad \text{STRONG INVERSION}$$

Calculation

Given that the *form* of the equation for the *depletion-layer width* in an MOS diode is the *same* as for an abrupt one-sided junction, calculate the maximum depletion-layer width in a p–Si substrate MOS diode where $N_A = 5 \times 10^{15}\,\text{cm}^{-3}$.

The form of the equation we're looking for is given by (6.30):

$$W = \sqrt{\frac{2\varepsilon_{Si}V_{bi}}{qN_D}} \qquad (6.30)$$

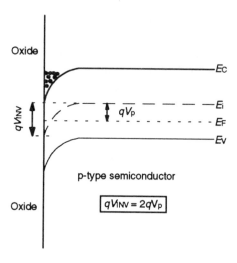

Oxide

qV_{NV}

qV_p

E_c

E_i

E_F

E_v

p-type semiconductor

Oxide

$$qV_{NV} = 2qV_p$$

Figure 9.9   Strong inversion.

Three changes need to be made to this equation. First, since we're concerned with the depletion region extending into p-material, the first change is

$$W = \sqrt{\frac{2\varepsilon_{Si}V_{bi}}{qN_A}}$$

Next, the potential acting is *not* the built-in junction potential but, rather, the *inversion potential*, so we replace V_{bi} by $2V_p$ (for the strong inversion condition):

$$W = 2\sqrt{\frac{\varepsilon_{Si}V_p}{qN_A}}$$

Finally, since we're in strong inversion the depletion-layer width will be the maximum depletion-layer width W_m:

$$W_m = 2\sqrt{\frac{\varepsilon_{Si}V_p}{qN_A}} \quad \text{DEPLETION-LAYER WIDTH, STRONG INVERSION}$$

The preceding *derivation* allows us to calculate V_p:

$$V_p = \frac{kT}{q}\exp\frac{N_A}{n_i} = 25 \times 10^{-3}\ln\frac{5 \times 10^{15}}{1.4 \times 10^{10}} = 0.32\,\text{V}$$

So the maximum depletion-layer width W_m is

$$W_m = 2\sqrt{\frac{8.854 \times 10^{-14} \times 11.9 \times 0.32}{1.6 \times 10^{-19} \times 5 \times 10^{15}}} = 0.41\,\mu\text{m}$$

I must point out a simplification I've made concerning the MOS diode before we carry out the next derivation. The MOS diode I've discussed is *even more ideal* than those usually considered. The MOS diode above has flat bands when the bias on the top electrode is *zero*. This is *not* generally true. Because the top electrode material often differs from the semiconductor on the other side of the oxide, there is an extra potential to consider, which I haven't done. This has nothing to do with charges in oxides or surface states as I mentioned earlier. It's due to the different *work functions* of the top metal electrode and underlying semiconductor. I really don't want to get into this stuff in this introductory book, but you can read it up in any of the slightly more advanced texts. In practice, it's easy to take into account (you simply add on a constant flat-band potential V_{FB} where necessary), but there's not enough payback for a full explanation involving work functions.

Derivation

Derive an equation for the *threshold voltage* of an ideal MOS diode.

When we are at threshold (have formed an inversion layer) the voltage on the top metal electrode (V_T) will equal the sum of the band-bending potential (V_{INV} or $2V_p$) and the potential due to the uncovered depletion-layer (width W_m) charge. By relating the depletion-layer charge to a voltage (potential) via the capacitor equation (6.59) we should be able to get the threshold voltage.

The discussion so far says

$$V_T = 2V_p + V_{DEP}$$

We already have an equation for V_p (equation (6.8)) so we just have to find an equation for V_{DEP}. The uncovered charge per unit area in the depletion region is simply

$$Q_{DEP} = -qN_AW_m$$

and if we want we can substitute for W_m using our earlier derivation:

$$Q_{DEP} = -2\sqrt{\varepsilon_{Si}V_pqN_A}$$

If we are considering what voltage *on the top metal electrode* gives rise to this charge *in the semiconductor*, then the relevant *capacitance* to consider is the *oxide capacitance* (think a little about this one). As an equation,

$$V_{DEP} = \frac{Q_{DEP}}{C'_{OX}}$$

or,

$$V_{DEP} = -\frac{2}{C'_{OX}}\sqrt{\varepsilon_{Si}V_pqN_A}$$

We can tie all these results together to get

$$V_T = 2V_p - \frac{2}{C'_{OX}}\sqrt{\varepsilon_{Si}V_pqN_A}$$

Finally, to *try* and avoid sign confusion, we know that the depletion-layer charge (uncovered acceptors) is negative, so we can write (treating all quantities as positive numbers)

$$V_T = 2\left(|V_p| + \frac{1}{C'_{OX}}\sqrt{\varepsilon_{Si}V_p q N_A}\right)$$

It should be clear that the oxide capacitance is the capacitance *per unit area*:

$$C'_{OX} = \frac{\varepsilon_{OX}}{d_{OX}}$$

where

ε_{OX} is the oxide permittivity (use a relative permittivity of 3.9 for silicon dioxide),

d_{OX} is the oxide thickness.

Calculation

Calculate the threshold voltage for an ideal p–Si MOS diode with $N_A = 5 \times 10^{15} \, \text{cm}^3$, and $d_{OX} = 120 \, \text{nm}$.

The previous calculation using the same N_A gave us

$$V_p = 0.32 \, \text{V}$$

and

$$W_m = 0.41 \, \mu\text{m}$$

From the above derivation we have

$$V_T = 2V_p + \frac{Q_{DEP}}{C'_{OX}}$$

Expanding gives us

$$V_T = 2V_p + \frac{q N_A W_m d_{OX}}{\varepsilon_{OX}}$$

Simply substitute the values to get

$$V_T = 2(0.32) + \frac{1.6 \times 10^{-19} \times 5 \times 10^{15} \times 0.41 \times 10^{-4} \times 1.2 \times 10^{-5}}{3.9 \times 8.854 \times 10^{-14}} \quad \text{volts}$$

Get out the calculator:

$$V_T = 0.64 + 1.14 = 1.78 \, \text{V}$$

9.4 MOSFET

The MOS diode structure that we've just studied has now been incorporated into a MOSFET as shown in Figure 9.10. What I've called the top metal electrode (for the MOS diode) is now called the *gate* in the MOSFET. This particular MOSFET is an *n-channel enhancement-mode* device and we'll see why it's this *type* in a moment. Note that if there is no bias applied to the gate electrode, then we have two n$^+$ regions (called the source and the drain) separated by a p-region, giving us an equivalent circuit of two back-to-back diodes, as we've seen before in the bipolar transistor. The *gate* in the *MOSFET* is like the *base* in the *bipolar* transistor. It is the controlling electrode, controlling much bigger currents flowing in the source–drain circuit (compare with the current flow in the bipolar emitter–collector circuit). How does it work? Figure 9.11 shows the MOSFET connected up to two voltage sources, one for the drain–source circuit, the other for the gate biasing. With no gate bias we know what happens. The source is isolated from the drain, and no current flows (except for a very small reverse leakage current). If we now apply a positive bias to the gate, and increase this bias above the *threshold voltage*, we shall form a conducting electron inversion layer (the *n-channel* part of the device name) under the gate. Note that the inversion layer in the MOSFET will form much more easily than in the MOS diode. The reason for this is that the MOSFET has two nearby highly doped n$^+$ regions, which are able to supply plenty of electrons to the inversion layer during its formation. We don't have to rely on thermal (or optical) generation only. The *inversion layer* will effectively connect the source to the drain, allowing current to flow in the source–drain circuit. Increasing the gate bias will put more charge into

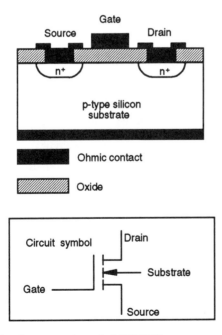

Figure 9.10   n-channel enhancement-mode MOSFET.

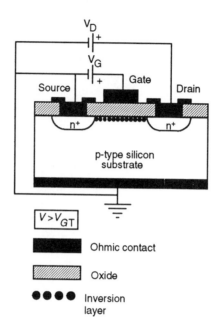

Figure 9.11   n-channel enhancement-mode MOSFET.

the inversion layer and more current will be able to flow in the source–drain circuit. That is why this is an *enhancement-mode* device. Increasing positive gate biases increase the source–drain current, and no gate bias means no source–drain current. The *depletion-mode* MOSFET we shall look at later *conducts* with *no gate bias*, and it requires a gate bias to *turn off!* Note that in Figure 9.11 the drain is biased positively with respect to the source. The source is therefore a source for *electrons* (not surprising since it is highly n-type doped) and *electron current* flows from source to drain (so *conventional current* flows from drain to source).

It is clear that we can use the MOSFET as a switch (via the gate bias), and it is also clear that we have considerable current gain (virtually no current flows into the gate since it sits on an insulator). Consequently, we also have considerable power gain. The device is behaving quite like the bipolar transistor we've just covered, except that the third terminal (the gate in this case) takes virtually no current. If the MOSFET *behaviour* is similar to the BJT we can expect a similar *I–V* output characteristic. Figure 9.12 shows the *I–V* characteristic for our n-channel enhancement-mode MOSFET. At first sight the *shape* of the characteristic *is* similar to the BJT (compare with Figure 8.18). But now look at the regions where *saturation* is occurring. Current saturation in the MOSFET is occurring where we intuitively feel it should, where the current doesn't increase with increasing drain–source bias any more. Remember, this is *not the saturation region* for the BJT. The BJT/MOSFET similarities are only surface similarities.

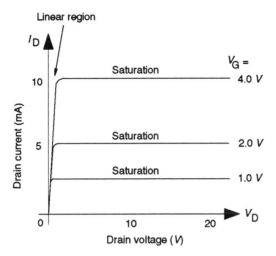

Linear region

Figure 9.12 *I–V* characteristic for n-channel MOSFET.

As we look a little deeper it is clear that these are two very *different* devices. We need some more physics to appreciate just how different these devices are.

9.5 MOSFET characteristics – qualitative

As is usual with a brand new (difficult) topic, I'll discuss what's going on with no mathematical formulation. Once we've got a reasonable picture of what's happening, we'll proceed with the maths.

The aim of this section is to qualitatively understand the MOSFET *I–V* characteristic of Figure 9.12. We've already seen that there are going to be problems because *saturation*, when discussing the MOSFET, is clearly completely different to *saturation* when discussing the bipolar transistor.

Figure 9.13 shows our enhancement-mode n-channel MOSFET operating in the *linear* region. We have a little more than the threshold voltage (V_T) on the gate so we have formed the conducting inversion layer. For *small* drain–source voltages the inversion layer behaves like a slab of ordinary resistive material ($V = IR$) so we get the expected *linear I–V* characteristic at low V_D.

Figure 9.14 shows the MOSFET at the point where current saturation is about to occur. What's going on here requires quite a bit of explanation. I've called the *drain voltage* at the point where current saturation is just starting to occur $V_{D(SAT)}$. If this voltage approaches ($V_G - V_T$) then there will be insufficient voltage drop across the oxide at the drain end of the device to maintain an inversion layer, and the inversion layer will disappear at this point! This condition is known as channel *pinch-off* for obvious reasons. Because the *source* is at *zero*

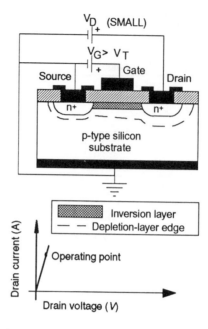

Figure 9.13   n-channel enhancement mode MOSFET in linear region.

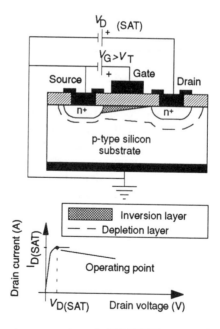

Figure 9.14   n-channel enhancement mode MOSFET in saturation.

potential we still have the *full* voltage drop (due to the gate potential) across the oxide at this end of the device, and hence the inversion layer is *unaffected* here. Assuming a linear voltage drop *along* the length of the inversion layer, it is clear that the inversion-layer thickness (zero at the drain end, maximum at the source end) will vary *linearly* as shown in the diagram. Note that a *depletion region* exists in the p-semiconductor near the drain contact (because of the increased positive drain potential), even though the *inversion layer* has pinched-off at the drain. What is not so obvious is why a current ($I_{D(SAT)}$) *still* flows with the channel pinched-off at the drain. I will try to explain why this current still flows after we've seen what happens when the drain voltage (V_D) is increased even further.

Figure 9.15 shows the MOSFET when $V_D > V_{D(SAT)}$. The pinched-off region moves towards the source end of the MOSFET, but the *potential* at the pinched-off point remains the same. We now have a small region between the end of the inversion layer and the drain without any free carriers, a depletion region. *Although this depletion region exists, it does not block the flow of electrons from the source to the drain.* The comparison that is often made is that we *also* have the flow of carriers across depletion regions in the p–n junction, and the bipolar transistor. Carrier flow *is not* blocked. Let's look in a little more detail at the situation we have in the MOSFET. The depletion-region length in the MOSFET is usually very small, which means that the electric field across it is very large. The electric field is acting to accelerate electrons from the pinched-off end of the inversion layer *into* the drain (the source is at zero potential and the drain is at some positive potential). *The electric field acting on the electrons is so large that they very quickly reach the saturation drift-velocity.* Go back to our early basic equation for current:

$$I_d = nAv_d q \tag{4.2}$$

If the drift velocity has saturated ($v_d = v_{d(SAT)}$) then the current has saturated ($I_d = I_{d(SAT)}$) provided n (the electron density) stays constant. The electron density we are concerned with is the density reaching the pinch-off point from the source. The electron density reaching the pinch-off point is governed by the *potential* at the pinch-off point, which as we've seen above remains constant. Tying all these facts together shows us that once the drain voltage has reached, or exceeded $V_{D(SAT)}$, the current *saturates* and remains constant at $I_{D(SAT)}$ (the saturated *drain* current).

You may still be unhappy with the carrier density reaching the pinch-off point being a constant due to the potential at the pinch-off point being a constant. We can crudely see that this is in fact OK by going back to the basic current equation yet again:

$$I_d = nAv_d q \tag{4.2}$$

Replacing the drift velocity by mobility times electric field:

$$I_d = nA\mu_n E q \tag{9.7}$$

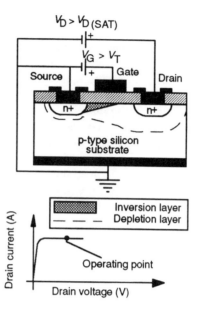

Figure 9.15 n-channel enhancement mode MOSFET beyond saturation.

where

    μ_n is the electron mobility,
    E is the electric field across the inversion layer.

Replacing the electric field across the inversion layer by the potential drop across the inversion layer and the inversion-layer length gives

$$I_d = \frac{nA\mu_n V_{po} q}{L} \qquad\qquad (9.8)$$

where

    V_{po} is the potential at the inversion-layer pinch-off point (so the potential drop across the inversion layer is V_{po} since the source potential is zero),
    L is the length of the inversion layer.

This equation shows that if V_{po} is a constant, then so is I_d. It also shows that if L decreases (which it *does* as V_D increases), then the drain current will increase (so we won't have a *constant* saturated drain current). This is what happens when you start looking at things in a little more detail; they always get complicated! To un-complicate things we tend to make approximations! *If* the inversion layer length L is *very large* compared to any *changes* in length due to V_D (i.e. *long* gate lengths), then the drain current will not appear to rise and we will have a (constant) saturated drain current. If on the other hand you're making very fast MOSFETs, then you're making devices with *very short* gate lengths (to reduce the transit time across the MOSFET). It is therefore not surprising that in these *short-channel*

MOSFETs you *will* see an increase of drain current (I_D) with increasing drain voltage (V_D). By *short* we usually mean MOSFETs with gate lengths of 1 μm or less.

I appreciate that even a qualitative explanation like that given above is a little difficult to follow. That being the case I won't go overboard with the maths. We'll just look at the *minimum* that you should know for this device.

9.6 MOSFET characteristics – quantitative

We will first consider the n-channel enhancement-mode MOSFET operating in the *linear region* (V_D small). For gate voltages above the threshold voltage an inversion layer forms and the inversion-layer charge per unit area is

$$Q_n = -C'_{OX}(V_G - V_T) \quad \text{for} \quad V_G > V_T \tag{9.9}$$

For completeness:

$$Q_n = 0 \quad \text{for} \quad V_G < V_T \tag{9.10}$$

Now consider Figure 9.16 for the MOSFET operating in the linear region. The channel resistance will decrease as V_G increases because Q_n increases. If the z-axis is *into* the plane of the paper, and the gate (channel) length is L, then

$$I_D = A \sigma E = (zx)nq\mu_n E \tag{9.11}$$

where σ is the channel conductivity (S/m), and we have used equation (4.14).

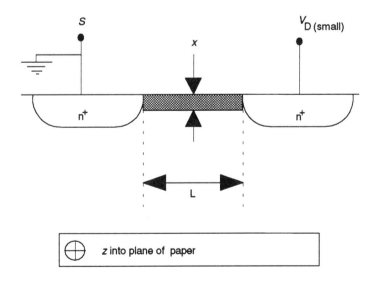

Figure 9.16 n-channel MOSFET in linear region.

Substituting the inversion-layer charge per unit area gives

$$I_D = z(-Q_n)\mu_n E \tag{9.12}$$

Finally, using equation (9.9) and substituting for the electric field along the length of the inversion layer gives

$$I_D = \frac{z\mu_n C'_{ox}(V_G - V_T)V_D}{L} \tag{9.13}$$

for the drain current in the linear region.

From equation (9.13) we can calculate the *channel conductance* in the *linear region*. The channel conductance g_D is defined as

$$g_D = \left.\frac{\partial I_D}{\partial V_D}\right|_{V_G} \tag{9.14}$$

Performing the differentiation on equation (9.13) yields

$$g_D = \frac{z\mu_n C'_{ox}(V_G - V_T)}{L} \quad \text{LINEAR REGION} \tag{9.15}$$

If we now make V_D large, we will start to pinch-off the conducting channel as in Figure 9.17. Because the inversion-layer charge is varying linearly along the channel length, we will use *an average value* of Q_n for the derivation of drain saturation current. Assuming saturation current is flowing, then

$$I_{D(SAT)} = z(-Q_n)_{AVERAGE}\mu_n E \tag{9.16}$$

Substituting for Q_n(average) and E (as before) gives

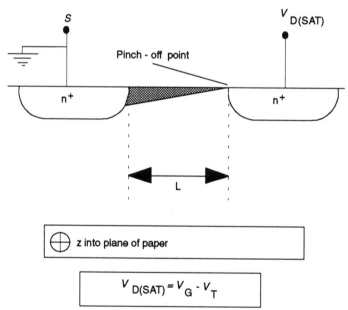

Figure 9.17   n-channel MOSFET at pinch-off.

$$I_{D(SAT)} = C'_{OX} \frac{(V_G - V_T)}{2} \mu_n \frac{(V_G - V_T)}{L} \qquad (9.17)$$

So, a slight rearrangement gives

$$I_{D(SAT)} = \frac{z \mu_n C'_{OX} (V_G - V_T)^2}{2L} \qquad (9.18)$$

or, equivalently,

$$I_{D(SAT)} = \frac{z \mu_n C'_{OX} V^2_{D(SAT)}}{2L} \qquad (9.19)$$

In this saturation case the channel conductance as given by equation (9.14) is zero. This isn't surprising since equation (9.18) is independent of V_D. However, another quantity, called the *transconductance*, is a little more interesting. The transconductance in the saturation region is given by

$$g_{m(SAT)} = \frac{\partial I_{D(SAT)}}{\partial V_G} \bigg|_{V_D} \qquad (9.20)$$

Carrying out the differentiation on equation (9.18) gives

$$g_{m(SAT)} = \frac{z \mu_n C'_{OX} (V_G - V_T)}{L} \qquad (9.21)$$

Calculation

Given

  gate oxide capacitance $= 4 \times 10^{-4}\,F/m^2$,
  channel width $= 50\,\mu m$,
  channel length $= 5\,\mu m$,
  electron mobility in the channel $= 750\,cm^2/V\,s$,
  gate voltage $= 6\,V$,
  threshold voltage $= 1.5\,V$.

Calculate

(a) the saturated transconductance,
(b) the common-source voltage gain with a load resistor (R_L) of $25\,k\Omega$ if the common-source voltage gain is given by $-g_{m(SAT)}R_L$.

(a) Using our equation (9.21):

$$g_{m(SAT)} = \frac{50 \times 10^{-6} \times 750 \times 10^{-4} \times 4 \times 10^{-4} \times (6 - 1.5)}{5 \times 10^{-6}} = 1.35 \times 10^{-3}\,S$$

(b) The common-source voltage gain is

$$gain = -g_{m(SAT)}R_L = -33.75$$

9.7 MOSFET – depletion mode

Figure 9.18 shows the basic structure of both n- and p-channel depletion-mode MOSFETs. I will discuss the operation of the n-channel device (Figure 9.18(a)), and will leave you to work out the functioning of the p-channel depletion-mode device for yourself.

Referring to Figure 9.18 you will see that the depletion-mode device is made with a conducting channel between the source and drain. This is achieved by doping the region under the gate oxide during device fabrication. With a conducting channel already in place, the depletion-mode MOSFET will conduct between source and drain with *no* applied gate bias. The depletion-mode MOSFET is therefore a *normally-on* device (the enhancement-mode MOSFET is normally-off). For the n-channel device we will need to apply a *negative* gate bias to deplete the (doped) n-channel and turn the device off.

So far the operation of this transistor should be quite clear: it's really just the converse of the enhancement-mode device. The only thing that needs a little explanation is the channel pinch-off condition with increasing drain voltage. Just like the enhancement-mode device, the channel in the depletion-mode MOSFET *will* pinch-off with large V_D. Figure 9.19 shows what is occurring around the source and drain regions in more detail. With no gate bias, and a large (positive) drain bias applied, the depletion region will appear as in the diagram. A large

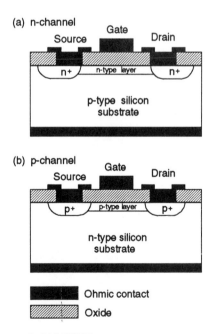

Figure 9.18 Depletion-mode MOSFETs.

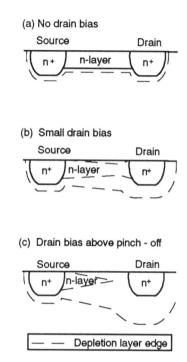

Figure 9.19 Depletion-mode MOSFET at pinch-off.

depletion region will form around the drain end of the device since we have a reverse biased n^+p diode where the p-part of the diode is the transistor substrate. We get the expected voltage drop in going from the drain to the source (which is at zero potential), so the depletion-layer thickness will diminish from drain to source. Note that the n-channel is already beginning to pinch-off at the drain end as the depletion region begins to extend into this region. For even higher drain biases the depletion region will extend *through* the n-channel near the drain end, and the channel will pinch-off.

As it is fairly difficult trying to keep all these different device types (and their behaviour) in your head at one time, I've produced Figure 9.20, which shows all four types of MOSFET, and their typical *I–V* characteristics.

9.8 MOSFET scaling

In the quest for higher-speed devices, and for more devices in a given area of silicon, it is clear that it is appropriate to make devices as small as possible. However, if other factors are taken into account, small devices may cause more problems than larger ones. The purpose of this section is to see what problems

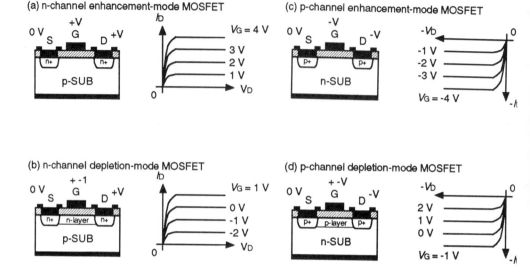

Figure 9.20   n-channel enhancement-mode MOSFET.

may occur as MOSFETs are reduced in size, and what general (scaling) rules are used to govern the size reduction.

We would like the channel length to be short to cut down the carrier transit time from source to drain. Figure 9.21 shows the source and drain regions for a long-channel and a short-channel MOSFET. Look at the effect of the depletion-layer edge. For the long-channel device the channel width taken up by the depletion layer is not significant. For the short-channel device we can immediately see a problem! When the sum of the source and drain depletion-layer widths equals the channel length, the depletion layers merge and we get *punch-through*. The gate no longer controls the channel current and MOSFET action ceases. This punch-through problem for short-channel MOSFETs is a major device limitation. By now you should have built up enough background knowledge to see how to partially alleviate this problem. If the substrate doping is *increased*, the depletion-region thicknesses around the source and drain will be reduced.

Another problem encountered in trying to make small MOSFETs is the effect of *hot electrons*. As device dimensions are lowered, electric fields will tend to become very large (unless potentials can be similarly lowered), and these large electric fields can accelerate carriers to high energy (we usually think of hot *electrons* because of their higher mobility). In fact, the carriers can be so energetic (hot) that they get accelerated into the gate oxide where they will reside as excess

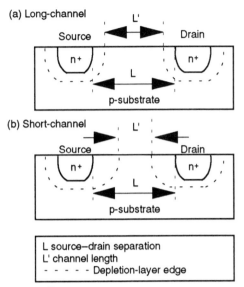

Figure 9.21 Short-channel effects.

charge in the oxide. Remember that in *real* MOSFETs we need to take this oxide charge into account as it affects the flat-band voltage, and hence the device *I–V* characteristics. I hope you can see the problem looming. If we are generating hot carriers that are getting injected into the gate oxide, then the device characteristics are going to change with operating time! This is not a very useful device.

From the above discussion, a logical way to scale MOSFETs would be to reduce *all dimensions and voltages* by some scaling factor, $\varkappa$, so that any electric fields would be of the same order as for a long-channel device. Before we apply these scaling ideas think about things that *may not* scale which could upset the scaling results. Remember, it's all well and good testing a scale model of a high-performance aircraft in a wind tunnel, but the results you get may not completely tally with reality because you didn't scale down the size of the air molecules as well! So keep in mind when we carry out this *simple* scaling exercise that we're not scaling down charges in oxides, work function differences between materials, defect densities in materials, and any other physically unscalable parameter you can think of.

Let's now reduce all device dimensions and operating voltages by $\varkappa$ and see what effect this has on some MOSFET parameters. First the oxide capacitance per unit area is

$$C'_{OX} = \frac{\varepsilon_{Si}}{d_{OX}/\varkappa} = \varkappa C'_{OX} \ \text{F/cm}^2 \tag{9.22}$$

The gate oxide capacitance is

$$C'_{OX}A = (xC'_{OX})\frac{L}{x}\frac{z}{x} = \frac{C'_{OX}A}{x} \text{ F} \qquad (9.23)$$

where

L = the gate length,
z = the gate width.

For the drain saturation current we get

$$I_{D(SAT)} = \left(\frac{z}{x}\right)\frac{\mu_n x C'_{OX}(V_G - V_T)^2}{2L/x} = \frac{I_{D(SAT)}}{x^2} \text{ A} \qquad (9.24)$$

But for the drain saturation current *density* we obtain

$$J_{D(SAT)} = \frac{I_{D(SAT)}}{A} = \left(\frac{I_{D(SAT)}}{x}\right)\left(\frac{x^2}{A}\right) = xJ_{D(SAT)} \text{ A/cm}^2 \qquad (9.25)$$

We can estimate the highest frequency of operation of our MOSFET in the following way. If the electrons cross the gate at the saturation drift velocity (v_{sat}), then the transit time will be

$$t_{tr} = \frac{L}{v_{sat}} \qquad (9.26)$$

Hence, the maximum frequency of operation (ignoring factors of order π) will be

$$f_T = \frac{v_{sat}}{L} \text{ Hz} \qquad (9.27)$$

So, applying our scaling laws gives

$$f_T = \frac{v_{sat}}{L/x} = xf_T \text{ Hz} \qquad (9.28)$$

The (current times voltage) *power dissipation* will scale as $1/x^2$.
The power-delay product (IV/f_T) will scale as $1/x^3$.
The power dissipation per unit area will yield a scaling factor of unity.

If you go through the above scaling factors the only one that causes us concern is the saturation current *density*. The saturation current density is *multiplied* by the scaling factor so we have to be aware of the *electromigration* problem (beginning of Chapter 7) in these scaled-down devices.

9.9 JFET

The junction field-effect transistor (JFET) is another unipolar, drift dominated, voltage-controlled transistor. A schematic of an n-channel JFET is shown in Figure 9.22. For *high speed* we want to move electrons around, hence the n-channel is appropriate, and we want the distance between the source and the drain to be as small as possible. The JFET is simply a piece of resistive silicon

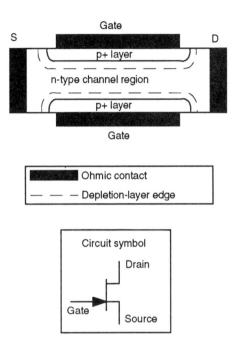

Figure 9.22   n-channel JFET (schematic).

whose channel conductance can be altered by the application of a (reverse) bias to
the gate terminal. The input impedance of the JFET is going to be *high* because
control is via the *reverse biased* gate junction; hence current flow is just the
reverse leakage current of a p–n junction. Before applying *any* gate bias let's look
at channel pinch-off with increasing *drain* bias (just as we did with the
MOSFETs).

Figure 9.23 shows how the depletion region near the drain end of the JFET
increases with increasing positive drain potential. This situation is very much like
the one we've already covered in the MOSFET. With a large positive potential on
the drain, and a zero potential on the source, we will have a large depletion layer
in the drain region of the JFET channel, tailing off into a much smaller depletion
layer in the source region. If we keep increasing the drain potential there will
come a point where the depletion regions at the drain end meet and we get
channel pinch-off. This channel pinch-off is also very similar to the MOSFET
case. Current flow *does not stop* at pinch-off, it continues, but it *is* saturated! So
just like the MOSFET, an increasing drain potential will eventually pinch-off the
conducting channel, and a saturation drain current ($I_{D(SAT)}$) will then flow.

Now let's look at the effect of the gate. Figure 9.24 shows the same JFET with
increasing reverse bias on the gate. As the reverse bias increases, the conducting
channel width decreases until eventually the depletion regions meet and the

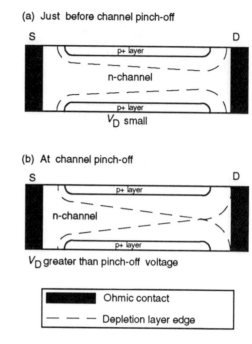

Figure 9.23   JFET channel pinch-off.

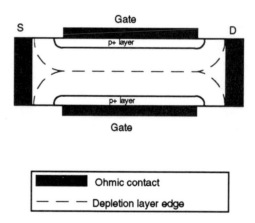

Figure 9.24   Large reverse gate bias.

channel is cut-off. At this point no electron current is able to flow from source to drain and the device is off. Hang on. I don't like that explanation one bit. You just said above that closing (pinching) off the channel using the drain potential didn't cut the current off; the saturation drain current $I_{D(SAT)}$ still flowed. Now

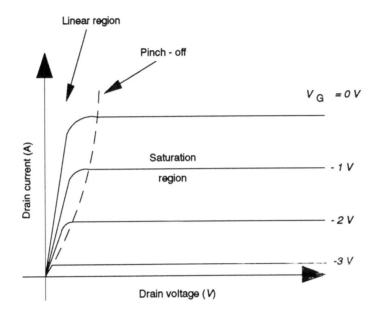

Figure 9.25 *I–V* characteristic for n-channel JFET

you're saying that closing (pinching) the channel off using the gate (instead of the drain) does stop current flow. You can't have it both ways!! Well, er, er, all right then, but do you see that the applied reverse bias on the gate means that pinch-off *will* occur at *lower* drain voltages because the channel width has been narrowed a bit by the gate depletion layer so *less* drain volts will pinch-off the channel. Yes, yes I *do* see that. Stop trying to cloud the issue. You said the reverse biased gate pushing the depletion region across the channel cuts the current off, but that the positive drain potential pushing the depletion region across the channel *doesn't*!

We'd better take a break at this point and look at the *I–V* characteristic of a *real* device to see what actually *does* happen. Figure 9.25 shows the *I–V* characteristic of an n-channel JFET. See that current saturation (pinch-off) *does* occur at lower drain voltages with increasing gate reverse bias (we expected that). Note also that a sufficiently large reverse gate bias *will* turn the device off! That's really strange. What's going on here? *The big mistake is to attribute the device behaviour to whatever the depletion regions are doing.* The depletion regions are the 'second order' effect. *The carriers are concerned with what the electric fields and potentials are doing, not what the depletion regions are doing!!* The shape and form of depletion regions *often* give a good indication as to what electric fields are present, but following the schematic pictures too closely can cause you to come unstuck, as in this case. What the depletion-region pictures *do not* show you is that a reverse biased *gate* puts a *potential barrier* in the way of electron flow from source to drain! If we have a drain potential alone *there is no barrier* to electron flow from source to drain. Even with the channel pinched-off (by drain potential alone) the electric field is *still* from drain to source, *aiding* electron flow from source to drain. *This is not the case when we apply a sufficiently large* reverse gate bias. *A potential barrier forms blocking carrier flow.* I don't want to get any deeper into this topic because it soon gets *very* sticky, but you should now have the general outline of the situation.

Some of you may have felt that the above was well over-the-top, and that no explanation was really necessary here. I'm very happy for you. Life for you is not complicated, and you won't be cultivating any ulcers. For some of us, though, the above type of problem can cause considerable grief. To show you it is a *real* problem, I encountered it again only recently, when trying to devise a new type of transistor. I had an idea for a transistor that had a base that was *fully depleted*, but *punch-through* did *not* occur, because there was a *potential barrier* in the base. This seemed to go against common sense because all the pictures you usually see in the books show that a depleted base means a punched-through (i.e. useless) transistor. You have to be *very* careful. In this particular case my 'new idea' had already been discovered. It's called an inversion base transistor, and, yes, it *will* function as a transistor even though the base *is* fully depleted!

9.10 JFET equations

The aim of this section is to derive the I–V equation for the JFET. Figure 9.26 is the diagram we shall refer to in the derivation, drain current is flowing and there is a voltage drop along the channel (y-direction). The dimension z is into the plane of the paper. As we've seen before, the voltage drop along the device leads to a depletion width that decreases from drain to source. If the gate is highly p-type doped, and the semiconductor is n-type doped (n-channel device), then we have a p^+n junction formed by the gate and the channel. We have already derived the depletion width as a function of voltage for an abrupt one-sided junction (equation (6.61)). The form is

$$W = \left(\frac{2\varepsilon_{Si}V}{qN_D}\right)^{\frac{1}{2}}$$

(9.29)

where V is the applied *reverse* bias giving the depletion region, and I have neglected the small built-in potential.

If V_P is the pinch-off voltage required to just close the channel then substituting a for W in equation (9.29) gives

$$V_P = \frac{qN_Da^2}{2\varepsilon_{Si}}$$

(9.30)

where V_P is *negative* for an *n-channel* transistor.

Combining equations (9.29) and (9.30) gives

$$W = a\left(\frac{V}{V_P}\right)^{\frac{1}{2}}$$

(9.31)

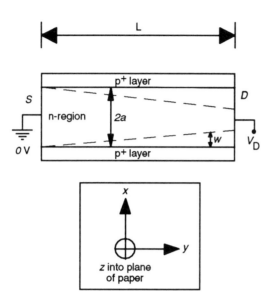

Figure 9.26 Diagram for *I–V* derivation.

Now the conducting height of the channel, w, varies along the length of the channel and is given by

$$w = 2(a - W) \tag{9.32}$$

The depletion-layer width is itself a function of the applied bias so we can combine equations (9.31) and (9.32) to give us an equation for the conducting channel height w, as a function of the applied bias V:

$$w = 2a\left[1 - \left(\frac{V}{V_{\text{P}}}\right)^{\frac{1}{2}}\right] \tag{9.33}$$

If we multiply equation (4.14) by area A, we get an equation relating current and conductivity σ,

$$I_{\text{D}} = -\sigma A \mathbf{E}_y \tag{9.34}$$

since I_{D} is in the negative y-direction. The drain current I_{D} is given in terms of the channel conductivity, the conducting channel area, and the electric field along the channel (y-direction). The channel area A is just wz, and we can rewrite the electric field as the variation in potential along the JFET giving

$$I_{\text{D}} = -\sigma w z \frac{\text{d}V}{\text{d}y} \tag{9.35}$$

We will get our current–voltage equation for the JFET by integrating equation (9.35) along the length of the device. For our n-channel device the applied gate bias is negative whereas the drain potential is positive. The integration will therefore be from $V = -V_G$ at $y = 0$ to $V = V_D - V_G$ at $y = L$.

$$\int_0^L I_D \, dy = -\int_{-V_G}^{V_D-V_G} \sigma w z \, dV \tag{9.36}$$

Substituting for w in equation (9.36) and rearranging gives

$$I_D = -\frac{2\sigma a z}{L} \int_{-V_G}^{V_D-V_G} \left[1 - \left(\frac{V}{V_P}\right)^{\frac{1}{2}}\right] dV \tag{9.37}$$

Performing the integration and noting that the constant of integration is zero since $I_D = 0$ when $V_D = 0$ gives

$$I_D = -\frac{2\sigma a z}{L} \left[V_D - \frac{2}{3}\frac{(V_D - V_G)^{\frac{3}{2}}}{(V_P)^{\frac{1}{2}}} - \frac{2}{3}\frac{(V_G)^{\frac{3}{2}}}{(V_P)^{\frac{1}{2}}}\right] \tag{9.38}$$

At saturation, I_D becomes $I_{D(SAT)}$ and $V_{D(SAT)} = V_G - V_P$. Substituting into equation (9.38) and rearranging gives

$$I_{D(SAT)} = -\frac{2\sigma a z}{L} \left[V_G\left(1 - \frac{2}{3}\left(\frac{V_G}{V_P}\right)^{\frac{1}{2}}\right) - \frac{V_P}{3}\right] \tag{9.39}$$

The signs are a real pain in this device! For this n-channel case, V_G is negative, V_P is negative, and V_D is positive. The only real way to see what's going on is with a calculation!

Calculation

An n-channel JFET has a channel height ($2a$) of $2\,\mu m$, a channel length of $10\,\mu m$, and a channel width of $250\,\mu m$. The channel doping density is $N_D = 3 \times 10^{15}\,cm^{-3}$. If the electron mobility in the channel is $1500\,cm^2/Vs$ find

(a) the pinch-off voltage,
(b) $V_{D(SAT)}$ for $V_G = -1\,V$,
(c) $I_{D(SAT)}$ for $V_G = -1\,V$.

(a) Use equation (9.30) to find the pinch-off voltage:

$$V_P = -\frac{1.6 \times 10^{-19} \times 3 \times 10^{15} \times (10^{-4})^2}{2 \times 8.854 \times 10^{-14} \times 11.9} = -2.28\,V$$

(b) We have the equation

$$V_{D(SAT)} = V_G - V_P$$

Substituting values in the above equation gives

$$V_{D(SAT)} = (-1) - (-2.28) = 1.28\,V$$

(c) Recall that

$$\sigma = nq\mu_n$$

Substitute this into equation (9.39) and then simply plug in the values to get

$$I_{D(SAT)} = -3.6 \times 10^{-3}\left[(-1)\left[1 - \frac{2}{3}\left(\frac{1}{2.28}\right)^{\frac{1}{2}}\right] - \frac{1}{3}(-2.28)\right]$$

Get out the calculator:

$$I_{D(SAT)} = -3.6 \times 10^{-3}[0.76 - 1 + 0.441] = -724\,\mu A$$

showing the current is in the negative y-direction. I did say that keeping track of the signs was a real pain!!

CHAPTER TEN

The semiconductor laser

Almost everyone in modern society must now be aware of the ubiquity of the semiconductor laser. This device has come from the research laboratory to our homes in just a few years, a truly incredible rate of commercialization. The compact disc recorder in your music centre, or car, or the small portable version you carry, all house a semiconductor laser. In the office, the laser printer also accommodates a tiny laser. The new optical communications systems which promise enormous data handling rates are based on fibre optic technology and the semiconductor laser. And to show just how far this device has come, it is now available (in red) as a viewgraph pointing aid! Coherent red light can be emitted, at power levels of a few milliwatts, from a solid-state device smaller than the full-stop at the end of this sentence.

I hope the introductory paragraph indicates the importance of this class of device. Having already entered our home as a 'workhorse' photon source, semiconductor lasers may well be at the heart of tomorrow's optical computing systems. But what is a laser? The acronym *LASER* stands for Light Amplification by the Stimulated Emission of Radiation. We shall look at *stimulated* emission and see how this differs from *spontaneous* emission, and why lasers need stimulated emission to work. But to answer what a laser actually is. A laser is a source of coherent (waves in step) monochromatic (single wavelength) light. This may not sound particularly interesting but we are simply not used to light sources being available that have such high power densities in such very narrow wavebands, and we can do a lot with such a source. The optical properties of laser sources allows us to make *holograms* (pictures vieweable in three dimensions) fairly easily, which is a highly interesting subject in itself, but unfortunately we will not be able to cover holograms here. The potential and applications of the semiconductor laser are enormous and they will continue to grow at an increasing rate for some years to come. This is why we must know how the device operates.

Figure 10.1 shows the three basic photon interaction mechanisms considered by Einstein when he was looking at the statistics of photons. The diagram has been drawn to illustrate a very general (rather than specific) situation. For example, E_1 could correspond to the valence band edge and E_2 to the conduction band edge in a (direct band-gap) semiconductor. I will only be considering direct band-gap semiconductors in this chapter, unless otherwise stated, because we are primarily concerned with photon emission. E_1 and E_2 can be any pair of radiatively connected energy-levels, where radiatively connected means that it is most likely that photons will be emitted when an electron drops from state E_2 to state E_1. The three optical processes shown are,

1 In Figure 10.1(a), the *absorption* of a photon of energy $(E_2 - E_1) = h\nu_{21}$, which promotes an electron from energy-level E_1 to energy-level E_2.

2 In Figure 10.1(b), the *spontaneous emission* of a photon when an electron falls from energy-level E_2 to energy-level E_1 emitting a photon of energy $(E_2 - E_1) = h\nu_{21}$ in the process. Physically it looks as if an electron simply drops (simultaneously) to a lower energy-level, with the emission of a photon. In reality there has been an interaction of the electron with the mass of virtual photons present in the vacuum (fortunately we don't need to get involved with such deep physics for a basic understanding of the LASER).

3 In Figure 10.1(c) we see a most interesting case, that of *stimulated emission*. We need this process to dominate over spontaneous emission if we want to make a LASER. In the spontaneous emission process a photon of the right energy, $(E_2 - E_1)$, has come along and found an electron in the upper energy-level state E_2. Because the photon has just the right energy it can cause the electron to drop to the lower energy-level (stimulated emission), but the incoming photon is not absorbed during the process. The incoming photon behaves like a catalyst. It causes the transition, and then carries on as it was before the interaction! The electron, however, makes its transition to the lower energy-level E_2 and emits a photon of energy $(E_2 - E_1)$, the same energy as the incoming photon. By this process we have obtained two photons of the same energy (frequency/wavelength) from the one incident photon. We thus have photon gain (a photon amplifier) and possibly the basis of a LASER. We are very fortunate that the stimulated photon also shares several other properties with the incoming photon; this is one of those very rare occasions where Nature actually works with us, rather than against us. Apart from the same energy/frequency/wavelength, the stimulated photon has the same polarization and phase as the incoming photon. A full understanding of why this is so requires some involved quantum mechanics. The stimulated photon also moves off *in the same direction* as the incoming photon rather than some random direction. For all intents and purposes the stimulated photon is *indistinguishable* from the incoming photon and this is most important if we are to build a LASER. Before we actually look at a simple LASER system, refer to Figure 10.2, which shows a material in which there are a lot more electrons in level E_2 than in level E_1. In this case when a photon of the right energy comes along it causes a great avalanche of electrons (by stimulated emission) to fall from E_2 to E_1, with the corresponding emission of a large number of photons, *all with the same $(E_2 - E_1)$ energy*! What I am suggesting is that there doesn't have to be a simple doubling of the photon density as in Figure 10.1(c). There can be an enormous photon gain

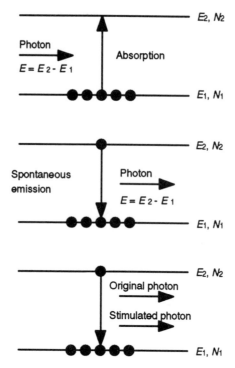

Figure 10.1 Optical processes.

provided we have a lot more electrons in a higher (radiatively connected) energy-level than the lower energy-level. The gain is therefore clearly going to be related in some way to the number density $(N_2 - N_1)$, the difference between the upper level density of electrons and the lower level density of electrons. But before we get too carried away, is this situation physically reasonable? For a material under normal thermal equilibrium conditions I think you can guess that the situation is *not* physically reasonable. For the simple case as we've considered so far we can immediately write down the population densities N_1 and N_2 by assuming a Boltzmann distribution

$$\frac{N_2}{N_1} = \exp -\frac{E_2 - E_1}{kT} \tag{10.1}$$

This equation is valid for a system in thermal equilibrium at temperature T. It tells us that the number density of electrons in the upper energy-levels falls off exponentially with increasing energy-level separation $(E_2 - E_1)$, and that there are always a lot more electrons in the lower energy-level!! Not very useful if we want stimulated emission to dominate. At this stage it is worth pointing out that if there are a lot more electrons in the lower energy-level than the upper energy-

(a)

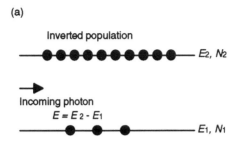

(b)

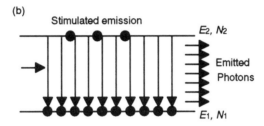

Figure 10.2   Photon amplification.

level, i.e. normal thermal equilibrium conditions, then the system will be a net photon absorber. So, systems in thermal equilibrium are photon absorbers. Clearly, how good an absorber a system is at the energy $(E_2 - E_1)$ is going to depend on some function containing $N_1 - N_2$ (recall that the *gain* would depend on $N_2 - N_1$, so it all hangs together so far). If we get the case where there are a lot more electrons in the upper energy-level than the lower energy-level then the system *can* show a net photon emission, but it is clearly *not* a system in thermal equilibrium. In fact, a system such as this is said to have an *inverted population (population inversion)* because the occupation of the energy-levels is inverted to the thermal equilibrium case. If this is so then it is clear that the thermal equilibrium equations do not apply in such cases. This does not stop people comparing the population inverted situation with the thermal equilibrium case using equation (10.1) and saying that a population inverted system appears to exhibit a *negative temperature*, since $N_2 > N_1$ and $E_2 > E_1$ so T must be negative. I don't think that this is particularly enlightening but I've mentioned it as you'll probably see it elsewhere. For the particular case where N_2 just equals N_1 we would have photon emission just balancing photon absorption, and for this case the material would appear *transparent* at a photon energy of $E_2 - E_1$! This is going to be a critical point, the point where if pushed one way the material is a net photon emitter (and possible LASER source), whereas if pushed the other

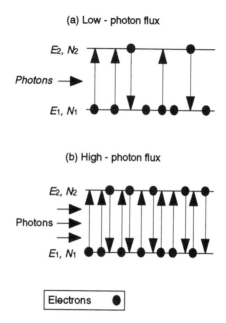

Figure 10.3 Attempt at population inversion.

way the material is a simple absorber. That is why, if you look at slightly advanced texts on LASERs you'll see that the conditions to achieve *transparency* are often mentioned. It's the point where the system is *about* to lase; it is on the *threshold* of lasing.

The above discussion should have made it clear that the key to making a laser is the ability to create a population inversion in a system. You might well assume that this is not going to be all that easy to achieve.

Let's start naïvely from square one and *guess* how we are going to produce an inverted population. What about hitting our simple two-level system very hard with lots of photons? We'll actually use a LASER to carry out this experiment. We'll use a LASER which has just the right energy output $(E_2 - E_1)$ and illuminate the material which has a pair of radiatively connected levels at this energy-level separation. Referring to Figure 10.3(a) and starting off with a low LASER power the material is seen to be absorbing (as expected) with a much higher concentration of electrons in E_1 than in E_2. As we increase the LASER power as in Figure 10.3(b) we naturally increase the 1 to 2 transitions but as the population builds up in level E_2 we also have more 2 to 1 transitions occurring due to the stimulated emission process *plus* the transitions due to the *spontaneous emission* which are going on *regardless* of the incoming photons. Note at this point that since the stimulated emission process requires photons to occur, the *rate* at which stimulated emission proceeds will depend on the *photon density*. In contrast, spontaneous emission is independent of the photon density.

So what's the best we can do? The unfortunate fact is that we cannot achieve population inversion in this system by this process. We can't even achieve a 50:50 population in E_2 and E_1 because the absorption and stimulated emission processes are going on at the same rate (see Appendix 2), *and* you've got the *added* loss from E_2 due to spontaneous emission. So a *simple* two-level electron system is not going to work for us. We shall see in a moment that the semiconductor laser *is* a simple two-level system, but we have an extra 'lever' in this system. We can *inject* carriers into semiconductors using junctions and thus 'bring in' the extra electrons or holes 'from outside' the system to get conditions right for lasing. You'll see how all this hangs together once we cover a more conventional laser.

OK, so a simple two-level system isn't going to work for us. What about three-levels? Figure 10.4 shows a hypothetical three-level system that could possibly work. Level 1 is the ground-state level. Most of the electrons are in here at room-temperature when the system is in thermal equilibrium. Level 2 comprises a broad band of energy-levels which allows me to use a (more convenient) broad-band radiation source (such as a Xenon flashlamp) to pump electrons up from level 1 to level 2. If level 2 were narrow I would need a narrow band of wavelengths (energies) for pumping, such as another LASER or a high power broadband source which is filtered down to give the right energy ($E_2 - E_1$). If the electrons can now move *quickly* from level 2 to level 3 (before they fall back to level 1) and if (yes we are multiplying up a lot of ifs here!!) level 3 is long-lived (also called *metastable*), then we could possibly build up the population of electrons in level 3. Perhaps if we hit the system quick enough and hard enough we could even get *more* electrons in level 3 than level 1. *If*, in addition, level 3 is radiatively connected to level 1, we could get net photon emission between levels 3 and 1. The photons to knock down the electrons (for stimulated emission) from level 3 to level 1 could come from some of the photons produced by the spontaneous emission process occurring between the same two levels.

I've just read over the above and I've convinced myself that such a combination of physical properties in a three-level system is not very likely to occur. I would say the possibility of such a fortuitous combination of events is even worse than remote! What makes the whole thing even more strange is of course that such a system *does* exist (in fact *many* such systems exist). It forms the basis of the first LASER made to operate, the *ruby laser*. Crystalline ruby is made from crystalline aluminium oxide (sapphire) which has a few fractions of a percent (by weight) of chromium impurity added. Funnily enough, crystalline ruby rods (boules) are made in much the same way as boules of single-crystal silicon (look up the Czochralski growth of semiconductor crystals in the standard texts). We can 'pump' the ruby rod with a broadband light pulse from a xenon flashlamp and the ruby will emit a pulse of light ($E_3 - E_1$) at 694.3 nm in the deep red. There is just one last thing to consider to turn our rod of ruby crystal into a LASER. What people mean when they say LASER is usually *laser oscillator*, a device that *produces* a narrow spectral bandwidth of photons at high-photon flux densities. Just as with any simple oscillator this means the device must exhibit gain *and* have

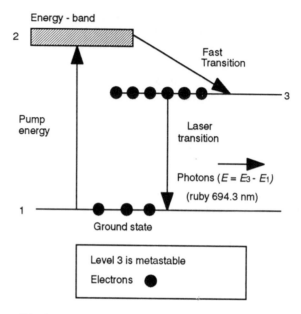

Figure 10.4   A possible three-level laser system.

positive feedback. Referring to Figure 10.5(a) we have our cylindrical ruby rod but the ends of the rod are antireflection coated so that they do not reflect any of the 694.3 nm radiation. If we now pump the crystal into population inversion (using the flashlamp) and then quickly let a pulse of 694.3 nm radiation pass through the rod, the light pulse will get amplified (by causing spontaneous emission from the upper laser level E_3) and a photon flux F_2 will be produced which is a much amplified version of the input flux F_1. Used in this *non-feedback* configuration we have a simple laser *amplifier*.

Optical feedback is pretty straightforward; just add mirrors. Figure 10.5(b) shows the laser *oscillator* configuration with a pair of mirrors parallel to the cut and polished ends of the ruby crystal. One mirror reflects 100% of the incident 694.3 nm light and the other mirror reflects most (but not all) of the incident radiation. The small fraction of a percent of light that is transmitted by the second mirror is in fact the laser output light. But where did this light come from? After optically pumping the ruby rod with a high-power xenon flash, spontaneous emission (noise) produced some 694.3 nm photons. Those spontaneous emission photons that were fortuitously produced going in the right direction (along the laser axis) were amplified and a small fraction was emitted as laser light at the output mirror. The reflected photons also produced more stimulated emission photons moving along the laser axis. You can see that a very large photon flux density builds up very quickly between the two mirrors, and the direction of travel is along the laser axis. Those photons that did not pass along the laser axis may be

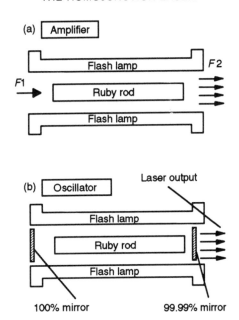

Figure 10.5   Laser amplifier and oscillator.

amplified a little, but then they are lost out of the side of the rod. Isn't it fortunate that the stimulated emission photons move off in the *same direction* as the incident photons so that we can produce an intense collimated beam of light along the laser axis!! Also note, just as in the simple electrical oscillator, our laser oscillator is triggered off in the first instance by *noise*.

I won't go into why a four-level laser system such as the neodymium YAG (neodymium ions in a crystalline yttrium aluminium garnet matrix) is even more efficient than a three-level system because we've covered enough material now to look at the laser system I really want to study, the semiconductor injection laser. If you're interested you can see why a four-level system is more efficient in Appendix 3.

10.1   The homojunction laser

Figure 10.6 shows the simplest structure for semiconductor injection lasers, the homojunction laser diode. The term *injection* simply means that the laser operates by injecting carriers across a junction, rather than being pumped by another mechanism (such as optical pumping or irradiation by high-energy electrons). You can see that its construction is very little more than an LED except that it has a pair of 'end mirrors'. In this instance the end reflecting faces are formed by cleaving (breaking) along the semiconductor crystal planes. The end 'mirrors' are thus aligned to atomic dimensions! The reflectivity from these faces is due to Fresnel reflection, which is caused by the refractive index difference between the semiconductor and the surrounding medium (air). The Fresnel reflection is given by

$$r = \left(\frac{n_1 - n_2}{n_1 + n_2}\right)^2 \qquad (10.2)$$

where

n_1 is the refractive index of air,
n_2 is the refractive index of the semiconductor.

You can see that since we square the result it doesn't matter if n_1 refers to the semiconductor or to the surrounding medium. Physically, this means that the percentage of radiation reflected is the same if we are dealing with a semiconductor/air interface or an air/semiconductor interface. Hence, for gallium arsenide (a common semiconductor laser material) with a refractive index of 3.66 we can form mirrors which will reflect about 33% of the incident band-gap radiation by simple cleaving. This percentage reflection is much less than that used for the ruby laser mirrors! We can get away with much lower reflectivity mirrors because the gain per unit length (gain coefficient) is much higher in the semiconductor laser.

We know how you get light out of a device such as that shown in Figure 10.6. Just as with the LED you forward bias the junction, and electron–hole

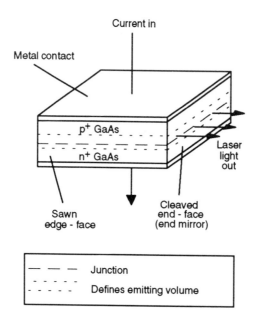

Figure 10.6   Homojunction laser diode.

recombination results in band-gap energy photons being emitted from the junction region. That's perfectly OK for the spontaneous emission radiation that we get from an LED but we need more than a pair of mirrors on an LED to make a semiconductor laser. We need the stimulated emission process to dominate. I don't like referring to population inversion when talking about the conditions for stimulated emission to dominate in semiconductor lasers because the situation is not the same as we considered earlier when there were only electrons. Now we have electrons *and* holes, so what must we do to get stimulated emission to dominate? Do we just need a large number of electrons in the conduction band? No, we can do that just by putting a large number of donors into the direct band-gap semiconductor. The Fermi-level shifts a bit and everything remains in thermal equilibrium. That's no good. The same thing goes for using acceptors to introduce a lot of holes. Just doping the semiconductor p-type and putting a lot of holes in the valence band is not going to make a laser either. But what about lots of electrons in the conduction band *and* lots of holes in the valence band at the same time, in the same region of semiconductor? Will that do? Yes! That will be fine because now there are plenty of electron–hole pairs available to recombine across the band gap to give out band-gap radiation (photons of energy $E_C - E_V$). Figure 10.7 shows the situation in detail. There is a slight problem with Figure 10.7(c). How do you get *lots* of electrons *and* holes together at the same time in the same piece of semiconductor? You can't dope highly p-type and n-type at the same time because all you end up with is n-type material if $N_D > N_A$ and p-type material if $N_A > N_D$. The way around this problem is to start with highly doped p-type material and then to inject lots of electrons into it using a p–n junction (just as in an LED) or to start with highly doped n-type material and then to inject lots of holes into it. Hence, the major physical difference (rather than structural) between the LED and the LASER is the much *greater* electron–hole pair density in the LASER (much higher injection levels). I've put this important topic into mathematical formulation in Appendix 4 because it's slightly too deep to include within this chapter but it's also far too important to leave out altogether.

We have seen a possible homojunction laser configuration in Figure 10.6 using GaAs as the photon emitting semiconductor. This laser clearly operates in the infrared. Why? Because the band gap of GaAs is 1.43 eV, which corresponds to photon emission at 0.87 μm. This wavelength lies in the infrared part of the spectrum. The light emission from the junction region will be asymmetric. Why? Because the minority carrier diffusion lengths are different for electrons in p-type GaAs and for holes in n-type GaAs, and it is these injected carriers, recombining with the majority carriers, that emit the photons. For clarification look at Figure 10.6. The finely dotted lines show the emitting volume. Because the electrons diffuse further into the p^+ GaAs before recombining, the light emitting volume is greater on the p-side of the junction than on the n-side.

Figure 10.8 shows the light output power of a laser diode as a function of the input drive current. Notice that below threshold, where lasing just starts, light

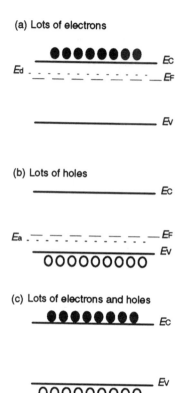

Figure 10.7   Carrier populations in a semiconductor.

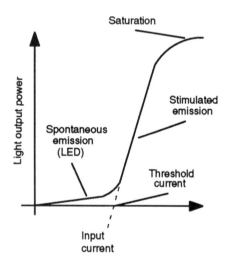

Figure 10.8   Ideal laser characteristic.

output is dominated by spontaneous emission. The laser diode under these conditions is acting like a simple LED giving out incoherent (waves *not* all in step) radiation! As the drive current passes the *threshold condition* stimulated emission begins to dominate and we get a rapid rise in light output power with drive current. Because of the nature of the stimulated emission process (the photons produced are all similar) the light emitted is coherent. Finally, note that at the highest drive currents the laser light output power begins to *saturate* (tail off).

There are a number of problems with the homojunction laser. The current required for the laser to reach threshold, the point where it just lases, is very high, about $5 \times 10^4 \, \mathrm{A \, cm}^{-2}$. This is partly due to the fact that we are not confining the carriers to a small volume; they are able to diffuse away from the junction region without constraint. Another problem is that we are not able to confine photons to the junction region either. The refractive indices of the material on either side of the junction are similar (it's the same material but with a different doping concentration) and this means we are unable to confine the photons. For a more efficient laser, i.e. one with a lower threshold current but capable of giving the same light output we desire the following:

(a) photon confinement,
(b) carrier (electron *and* hole) confinement.

Why does this help? Confining the carriers to a small volume pushes up the probability of their radiative recombination. Confining the photons to the same small volume pushes up the stimulated emission rate which is what we want for lasing. We are extremely fortunate that, again, Nature has decided to play along with us and allows us to have both conditions together in the double-heterojunction laser diode (DHLD).

10.2 The double-heterojunction laser diode

Figure 10.9 shows the basic structure of the double-heterojunction laser diode (DHLD). Note that just one simple change has given us *both* conditions (photon *and* carrier confinement) in one go. How has it done this? The *active layer*, which is the photon-generating region, is straight GaAs. The active layer, as you can see, has layers of AlGaAs above and below it. Since the band gap of AlGaAs is greater than GaAs we get carrier confinement to the GaAs layer, as shown in Figure 10.10(a). Because we have *different* materials systems on either side of a junction we have a *heterojunction* as we saw in the bipolar transistor section. The *discontinuities* in the conduction and valence band edges which lead to the carrier confinement are not of the same size, as you can see in the diagram. However, the size of the discontinuity *is* sufficient to give effective carrier confinement. What about photon confinement to the active layer? Here we have a bit of luck. Again looking at Figure 10.10(b) you can see that the refractive index of the active GaAs layer is *greater* than the refractive index of the surrounding AlGaAs

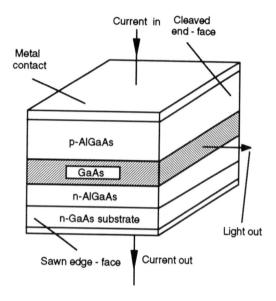

Figure 10.9   Heterojunction laser diode.

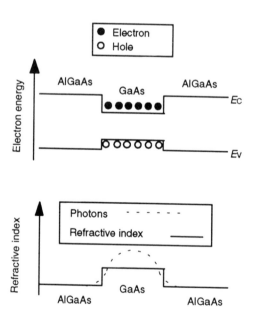

Figure 10.10   Carrier and photon confinement.

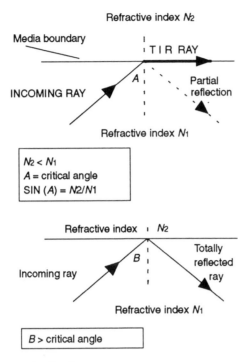

Figure 10.11   Total internal reflection.

cladding layers. These are the conditions required in order to get total internal reflection of the photons at the AlGaAs/GaAs interface and therefore confinement of the photons to the active layer. Figure 10.11 shows the conditions for total internal reflection. The active layer in a DHLD therefore acts like a *slab waveguide* and photons prefer to move in the active layer. Two very important points need to be considered:

1 The cladding layers simply provide a bigger band-gap material with lower refractive index. There is no need for photon production within the cladding layers (photon production takes place in the active layer, hence its name). This being the case *there is no need for the capping layers to be direct band-gap material.* This relaxes the material parameter requirements considerably and allows a degree of flexibility in producing lasers. The same material system can be used to produce slightly different wavelengths just by altering the material's composition slightly. Different laser wavelengths can also be produced by using active layer widths of varying size. These *quantum-well lasers* are discussed in Chapter 11.

2 *A bigger band gap does not necessarily mean a smaller refractive index.* Quite often, you will find that a bigger band gap *does* lead to a smaller refractive index, but this is not a law. This is a shame because it means that quite a few

promising laser systems are unsuitable for DHLDs, even though suitable materials of different band gap are available.

A crucial issue I haven't mentioned until this point is *lattice-matching*. A good-quality *heterojunction* requires that the crystal lattice spacing of the capping layer is almost exactly the same as the active layer. This explains the ubiquity of the AlGaAs/GaAs system, where almost perfect lattice-matching can be achieved. If the matching is not perfect then non-radiative recombination centres (among a host of other active sites) can exist at the heterojunction interface which can (and often do) make laser production impossible. Clearly, non-perfect lattice-matching can lead to various types of crystal *dislocations* at the heterojunction interface, and it is these, together with other crystalline irregularities, that lead to the problems. Good-quality lattice-matched interfaces also exist for a few other material systems such as GaInAsP/InP (which is used for long-wavelength communications lasers) and AlGaInP/GaAs (which is a system used to produce visible wavelength laser diodes). Note that with these quaternary (four-element) compounds you stand an even better chance of lattice-matching capping layers and active layers (by varying the atomic percentage of each element in the compound) since you have an extra free parameter (over the ternary, or three-element compounds), namely an extra element.

Finally, it's all well and good having a potentially excellent laser materials system, but you still need a good-quality, single-crystal starting layer, called a substrate, on which to grow your capping and active layers. The substrate lattice-constant has to be well matched to the capping layers or the capping layers will be full of defects which results in poor electrical quality. Sometimes the capping layer is not the first layer to be grown on the substrate, for a number of reasons, but in these cases we still require good lattice-matching throughout the multilayer device. The growth techniques typically used to deposit the laser diode semiconductors layers are molecular beam epitaxy (MBE), and metal-organic chemical vapour deposition (MOCVD). These growth techniques can be looked up in the standard texts. Going back to the substrates, there aren't really that many around which you can readily obtain to try out your new laser ideas. Of course there's silicon, gallium arsenide, indium phosphide, sapphire and a few others but then there's the question of substrate size availability. We like to be able to get large-diameter substrates so that we can make many circuits (lasers, microprocessors, or whatever) in one go. This 'mass-production' really brings down the commercial cost. Unfortunately, the substrates that are useful for lasers come in much smaller diameters than silicon, and silicon doesn't have the right lattice parameter for any present-day laser system. Silicon can be obtained as wafers (substrates) in excess of twelve inches (30 cm) in diameter! For GaAs four inches is just now becoming typical, with six inches being at the research stage. Other single-crystal substrate materials are similarly restricted in diameter making the cost per laser fabricated on them quite expensive. Let's summarize some of the requirements for fabricating a DHLD. You need the following:

1 A suitable single-crystal substrate material on which to grow the laser.

2 A suitable growth technique with which to grow the active and capping layers. This will require certain chemical compounds (such as metalorganics) with the right properties which may not exist for a newly proposed laser materials system!

3 A suitable materials system which gives a capping layer with a larger band gap and smaller refractive index than the active layer. The capping layer must also be well lattice-matched to the active layer, although it need not be a direct band-gap semiconductor.

4 A suitable doping source to provide us with p- and n-type cladding layers. This allows us to inject both electrons *and* holes into the active layer. Bipolar doping has been a substantial problem to those wishing to make *visible layers* using the 2–6 (II–VI) compound semiconductors, which include ZnSe, ZnTe, CdTe, etc. Unfortunately, these materials typically come as *either* p-type *or* n-type and type-conversion (to any reasonable extent) has not been possible. Reasons for this come under the heading of *dopant autocompensation* in more advanced texts, but in some cases it is most likely that it is a materials production problem (the material is full of some sort of defect which renders it highly p- or n-type and you cannot put enough dopant in to type-convert because it would go beyond the *solubility limit* of the dopant you're trying to put in). Having said that, scientists at the 3M Company (St Paul, MN, USA) have recently made the ZnSe system work! A lot of diligent research on a complex materials problem allowed them to come up with the first prototype 2–6 laboratory lasers. Their technique, which used MBE as the growth method, has demonstrated the feasibility of making that Holy Grail of laser physicists, *the blue injection laser!*.

5 A suitable metallization technology is required to put down the low resistance ohmic metal contacts (these contacts will carry HIGH current densities) on the laser materials. This again may seem trivial. Be warned, metal contact technology is a science (some would say black-art) in itself. The problem? If you want a good ohmic metal contact Nature wants to give you a metal-semiconductor rectifying contact, a Schottky diode. It's as simple (and frustrating) as that. The converse situation goes without saying. Those wanting to make Schottky diodes end up with ohmic contacts. You should know the name of the law that states this!

These are clearly the major points; many minor points need to be satisfied as well. But having produced a DHLD you are rewarded with a much lower *threshold current* for your efforts so your laser will put out the same light power for less electrical input power than the simple homojunction laser. This is due to the carrier *and* photon confinement to the active region.

In practice, the threshold current is *still too high* for many applications (especially in optical communications), even with the DHLD structure. For this reason the simple DHLD structure we have looked at is slightly modified in order to lower the necessary current for lasing (the threshold current) even further. One modification (many are possible) gives us the stripe laser (often called a gain-guided laser).

10.3   The stripe laser diode

Figure 10.12(a) shows one way to reduce the injection current into a DHLD. The
solution as you see is rather trivial. You simply constrain the lateral extent of the
current injection. Up until now the lasers we have looked at have all been *broad-
area* lasers, meaning that their lateral exent has been defined simply by the edges
of the laser diode. In the stripe laser diode we *deliberately define* the width of the
injecting contact by restricting its lateral extent in an oxide *window*. Since we are
now causing a smaller volume of active material to lase, it should come as no
surprise that the current requirements are reduced. Although this is also called a
gain-guided laser (since you only get the gain in the high-current-density regions)
it may as well be called a loss-guided laser since regions outside the high-density
region are not pumped, and are therefore lossy (absorbing).

Figure 10.12(b) shows another method of achieving the same result. In this
case instead of using an oxide to block the current path, ion damage converts
some of the semiconductor to intrinsic (insulating) material, with the same
resulting effect as using an oxide layer.

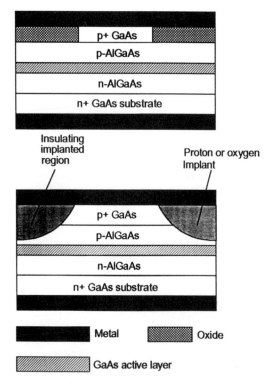

Figure 10.12   Stripe diode lasers.

Note that this gain-guided laser has not laterally confined the photons in the same way as a refractive index difference does. The lateral photon spread is restricted by loss in the unpumped active-layer regions and is therefore not very well defined. If, in addition to using a narrow injection stripe we also *completely surround* the active layer with lower index material, then we will also get *index-guiding*. Such a device is realized in the *buried-heterostructure* laser. Note that if we want high laser light output powers, we don't want such lateral confinement, and so we go back to broad-area lasers and other devices of a similar structure.

10.4 Index guiding

Figure 10.13 shows the structure of an index-guided laser, also called a *buried-heterostructure* laser, since the active layer is *buried* in material of lower refractive index. In this instance you can see that the active region is completely surrounded by material of lower refractive index. You clearly can't do much better than this for carrier and photon confinement. Look at the structure carefully. It's more a credit to the crystal grower and the device fabricator than anything else! The refractive index difference leading to photon confinement in structures such as that shown in Figure 10.13 is typically about 0.2. This refractive index difference

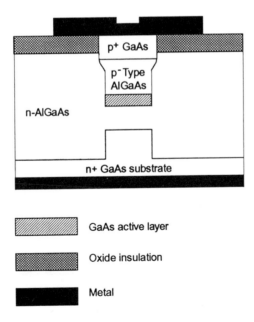

Figure 10.13 Buried-heterostructure laser end-view.

is more than two orders of magnitude greater than the refractive index difference caused by current injection effects (as in the gain-guided laser). This being the case, the beam shape is much better defined in the buried-heterostructure laser and it gives a much higher quality optical beam than an equivalent gain-guided laser. Other benefits include a low threshold current of about 10 mA and a higher modulation frequency (rate at which the light can be pulsed) of about 2 GHz. These latter properties make the buried-heterostructure laser a useful source in high bit-rate, long-haul optical transmission (fibre optic) systems. You will also find this type of laser in compact-disc players and laser printers.

10.5   Linewidth narrowing

Although the laser has a much narrower spectral distribution than the corresponding LED, as shown in Figure 10.14, it still isn't good enough for the very long-haul fibre optic transmission links at high bit rates. The problem is that the simple Fabry–Perot cavity we've been looking at up until now can support a large number of *modes*. I need to define a couple of new things here! A Fabry–Perot cavity for our purposes is a slab waveguide with mirrors at either end; in other words it's our laser cavity (active volume). Modes need a little more explanation. Look at Figure 10.15, which shows a pair of separated mirrors with some light waves travelling between them. The condition that light waves may travel easily between the mirrors is for each mirror to be at a node. This is exactly the same physical situation as for a piece of string stretched between two points. Pluck the string and it will vibrate; the ends of the string do not, these are the fixed nodes. However, the string is permitted to vibrate at a number of different frequencies between the two fixed points in a manner similar to that shown in Figure 10.15 for the light case. These different *allowed frequencies* are different

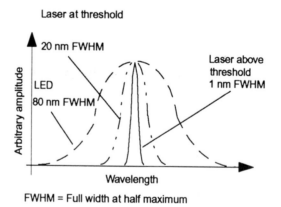

Figure 10.14   Laser and LED linewidths.

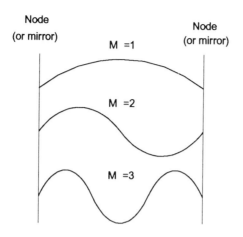

Figure 10.15 Standing waves.

allowed modes of the oscillating system. Clearly, the fixed points will support frequencies such that an integer number of half-wavelengths fit between the fixed points. We can write this mathematically as

$$\frac{(m\lambda)}{2} = L \tag{10.3}$$

where

    m is an integer,
    L is the cavity length (distance between the fixed points),
    λ is the wavelength.

We have assumed a refractive index of unity here. For a refractive index of n, divide the wavelength by n (and don't forget that the refractive index is a function of wavelength). This is looking a bit odd here. For our Fabry–Perot laser cavity we know that the light wavelength is pretty short, and by comparison the cavity length is pretty long, so aren't there a large number of allowed modes in our laser?

Calculation

For an AlGaAs laser with a cavity length of 100 μm, calculate the total number of modes that are theoretically supported, assuming unity refractive index. Is this number of modes supported in practice?

The first part is quite straightforward. A GaAs band gap of 1.43 eV (the active layer) corresponds to an emission wavelength of 0.87 μm (870 nm). Substitute

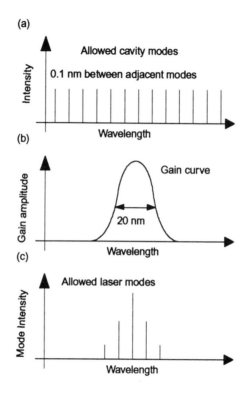

Figure 10.16 Allowed laser modes.

values into equation (10.1) to obtain $m = 230$ (to the nearest integer). This large number of allowed modes *will not* be supported in practice since only those frequencies that lie within the allowed laser linewidth (the gain curve) will be supported. I have shown the situation in Figure 10.16. In Figure 10.16(a) we see some of the large number of allowed modes as a function of frequency. In Figure 10.16(b) we see the laser light output intensity as a function of frequency which I am taking as the same thing as the gain curve, i.e. the range of frequencies over which gain occurs together with the magnitude of the gain. If we superimpose the gain curve over the total number of allowed modes we see the real number of modes that the laser cavity will support. In this example it comes down from a couple of hundred to just a few!

OK, so our Fabry–Perot laser cavity will support a few modes. What's the problem? I've still used a buried-heterostructure configuration so the threshold current is nice and low. I've now also chosen the GaInAsP semiconductor system to give me laser output at 1.55 μm where glass optical fibres have their lowest absorption (GaAs is no good here; it has the wrong wavelength and the glass fibres are too absorbing). What more do you want? This looks a pretty good system to me for long-haul fibre optic communications! It's almost there, but there's one last problem to solve. Our optical fibre is a dispersing medium. Different frequencies travel at *different speeds* down the fibre. This means that if you put a short pulse of light in at one end of the fibre, and this pulse is composed of several different frequencies, then they will travel at slightly different speeds along the fibre and will come out at the other end of the fibre at different times. Depending on the length of the fibre, the short pulse can end up being a long pulse (long enough for two consecutive pulses to overlap and thus destroy the information content) and this will cause problems. We've got to get rid of the unwanted side-modes (the modes on either side of the main mode) and just keep the main one at the gain maximum. As an aside, but just for completeness, the situation is made even worse by modulating (pulsing) the laser at high frequency. Under these conditions the power in the side-modes starts to increase and this makes the situation even worse. We now know the problem; so how do we fix it?

The problem is these side-modes. The reason they exist at all is because the Fabry–Perot cavity provides feedback (via the mirrors) which is equally good for all wavelengths (near the gain maximum). The point is that the optical feedback *is not* frequency-selective in the simple Fabry–Perot cavity. We need frequency-selective feedback! One way of doing this is by incorporating a diffraction grating into the laser structure. Unlike the reflecting *facets* at the end of the semiconductor laser cavity, the diffraction grating *does* provide frequency-dependent feedback. The grating can be implemented in two different ways, as shown in Figure 10.17, namely distributed feedback and the distributed Bragg reflector. In the distributed feedback system the grating lies along the whole length of the active laser medium, and only those wavelengths satisfying the grating equation will be supported. Don't worry too much about the detail here. It simply means that the grating only allows one wavelength to propagate along the laser axis. For distributed Bragg reflectors the grating is formed near the cavity ends and therefore distributed feedback does not occur in the central active region. The distributed Bragg reflectors in the unpumped end regions of the laser diode act as wavelength-dependent (only reflecting one wavelength) mirrors.

This last implementation of frequency-dependent feedback now gives us all the necessary properties for a laser diode to function efficiently in high bit-rate, long-haul fibre optic communication links.

10.6 The future

I have been loath in this chapter to put in absolute values, such as the size of lasers or the threshold currents for lasing because the technology is improving, and changing, at a phenomenal rate. Records are being broken almost monthly, and there's nothing more disgruntling than specifying some record-breaking technological achievements in a book, and then to pick it up just a couple of years

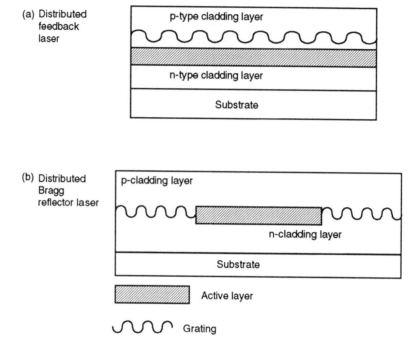

Figure 10.17   Feedback mechanisms.

later, when the 'achievements' are almost laughable! So what does the future hold?

In this chapter we have looked at laser diodes of 'normal' dimensions of perhaps $100\,\mu$m long and with constrained lateral dimensions of perhaps $10\,\mu$m with an active-layer thickness of the order of a micron (10^{-6}m). You will see in Chapter 11 that by making the active-layer thickness much smaller you can actually change the light-output wavelength slightly. You aren't stuck with laser light wavelengths governed by the active layer's band gap! These are called quantum-well lasers. We can go even further. If we also bring down the *lateral* dimensions enormously (to around $5\,$nm or so) we obtain *quantum-wire* lasers with considerable carrier and photon constraints in two dimensions instead of one. Amongst the many benefits this type of device will bring are extremely low threshold currents (microamps or less?) so that highly efficient, low-power light sources will be produced. And what about applications? We know that light oscillations occur at much higher frequencies than for radio or TV and, as a consequence, light has a much higher *information capacity* for transmission than radio or TV transmissions. It is in this area of information or data handling that photon systems will make a considerable impact. We are only just beginning to see the installation of optical data links across the oceans. Look at the rate of technological progress over the last decade accompanying the increased availability of information/data using computers linked by wire or radio. Now multiply that by at least a few hundred for optical data links and try to fathom the consequences. Now change your computer, because it's running far too slow at a lousy 50 MHz to handle all the optical data, to an all-optical computer running many thousands of times faster. At this point in time the outcome is unimaginable. Certainly, the impact on society should be at least as great as the development of the transistor. It's likely to be very much greater. And now look back at the transistor's impact over the last twenty years!!!

CHAPTER ELEVEN

An introduction to the quantum theory

Quantum theory, however taught, is always a shock to the system for a new student. This is unfortunate because a lot of 'forefront-of-technology' calculations are possible with only a rudimentary knowledge of quantum mechanics. If you can, I strongly advise you to do the following. Do not try to understand the mechanics of the theory at this stage. Use the mathematics that are presented in this chapter in order to understand the new 'low-dimensional devices' (LDDs) only. Virtually all the new quantum devices and quantum effects in semiconductor physics can be understood if you are well versed in just one quantum mechanical solution. The particular solution we are going to study is called 'the particle-in-a-box'. This may not sound very glamorous but the theory is easily understood and the applications are far reaching. For this reason, the way I am going to introduce the quantum theory is biased with just one aim in mind, to understand the physics and applications of the particle-in-a-box.

It is of course unfair to expect you to take things at face value and not look further than the brief introduction that will be given here. Please do look further, but also do not be put off by the deep philosophical and theoretical problems that appear when studying this subject. My approach will shield you from the deeper theory but the bravest among you will go further. As a challenge, consider the words of Richard P Feynman:

No-one really understands quantum mechanics!

That's quite a statement from a Nobel prizewinner whose PhD work was to produce a quantum theory by a completely different route to those already in existence. The two (equivalent) quantum theories already in existence were the *wave mechanics* of Schrödinger and the *matrix mechanics* of Heisenberg. Feynman's quantum theory was based on a *space–time* approach! We shall look at quantum theory based on the wave mechanics of Schrödinger because, conceptually, it's the easiest one of the three to grasp.

11.1 The wave–particle duality

As this book is not a historical treatise, and I have tried to keep everything up to date without continual reference to the past, some of you may need to refresh your memories, or do a little extra reading, if the experiments I mention in this section (but do not detail) are new to you.

It has been known since the beginning of this century that waves can exhibit particle-like behaviour. What is meant by this? Consider the *photoelectric effect* where light of sufficiently high frequency can eject electrons from the surface of some metals. Although we know that light exhibits wave behaviour (diffraction, interference), we do not need to consider wave behaviour at all to explain the photoelectric effect. The photoelectric effect can be described by considering light to consist of particles of a particular energy which can eject electrons from some solids. The *Compton effect* is where high-energy light photons (X-rays) collide with electrons and come out of the collision with lower energy. The Compton effect can be described totally in terms of classical particle collisions without any reference to waves. This is what is meant by waves exhibiting particle-like behaviour.

It now seems strange that nearly twenty years were to pass before Louis de Broglie considered the implications of the converse situation, that is, whether particles could exhibit wave-like behaviour. De Broglie's thinking was as follows. A photon of light of frequency v has momentum p given by the equation,

$$p = \frac{hv}{c} \tag{11.1}$$

This can be written in terms of the light's wavelength λ as

$$p = \frac{h}{\lambda} \tag{11.2}$$

since $\lambda v = c$. The wavelength of a photon is therefore given in terms of its momentum by

$$\lambda = \frac{h}{p} \tag{11.3}$$

Before we carry on with de Broglie's analysis there are a couple of points worth mentioning about the above equations. First, if you are unhappy about light having momentum, the physics of this can be looked up in any introductory text to special relativity. However, without resorting to mathematics, you should already be aware that light (photons) carry momentum. For those of you who like science fiction, recall the stories about solar sailing where vast sails of aluminized plastic drive small vessels by the pressure of sunlight. Although a fictional story at present, the idea is quite feasible. Many of you will have also seen an instrument built to demonstrate the 'pressure' of light. Remember the toy (radiometer) built around a light bulb where a black and white vane whirled around within the bulb?

Nearby was usually a light source to propel the vane. Although this is again quite feasible, the problem with these particular toys was that the vacuum was not quite good enough! If you looked carefully the black vane was being pushed away from the light, when it should have been the white vane. The reason for this is that the residual gas was being pushed away from the warmer black vane and as a reaction the vane rotated, but in the wrong direction!

Returning to de Broglie, he now considered the possibility that equation (11.3) need not just apply to light, but to material particles as well. The momentum of a particle is given by

$$p = mv \qquad (11.4)$$

where m is the particle mass and v is the particle velocity. And so the wavelength that might be associated with this particle is given by

$$\lambda = \frac{h}{mv} \qquad (11.5)$$

by using equation (11.3). This wavelength is called the particle's de Broglie wavelength. Note that the higher the particle's momentum, the shorter its de Broglie wavelength.

As stated earlier, I don't wish to study history in this book but in this instance the de Broglie story warrants it. The above analysis by de Broglie formed part of his PhD thesis, and clearly when his professors read this work they initially though that maybe de Broglie had been concentrating a little too hard on his topic! However, before having de Broglie committed they wisely decided to send the thesis off to Albert Einstein. A fortunate decision since Einstein confirmed that this was indeed good work!

Experimental proof of the wave behaviour of particles was given in 1927 by Davisson and Germer, who diffracted (a purely wavelike phenomenon) electrons from a single crystal of nickel. Today, electron diffraction is widely used in a growth technique called molecular beam epitaxy (MBE). A low energy electron beam is diffracted off a silicon or compound semiconductor slice before growth begins to check that a good single-crystal surface is present. This technique is called RHEED for reflection high-energy electron diffraction.

We have seen in the above discussion how particles can show wavelike behaviour and how waves can show particle-like behaviour. This wave–particle duality is fundamental to an understanding of quantum mechanics. Before proceeding to a study of quantum mechanics it is usual to reconsider the older atomic theories of Bohr and Thomson. I shall not do this but I will show an instance where classical theory fails to account for behaviour at the atomic level.

11.2 A failure of classical physics

In the classical picture of the atom, electrons orbit around the positively charged nucleus held in place by electrostatic forces. Figure 11.1 shows the situation for a

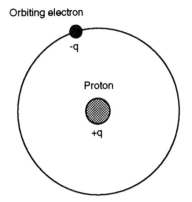

Figure 11.1   Classical hydrogen atom.

hydrogen atom. The single electron is classically orbiting the single proton, held in this orbit by an electrostatic force. Unfortunately, even classical theory wouldn't allow such a picture. An accelerating charge radiates energy according to classical (and modern) electromagnetic theory, and an orbiting electron is an accelerating electron. The accelerating electron would therefore radiate energy, the electron in losing energy would then move closer to the nucleus and its acceleration would increase. Figure 11.2 shows the physical situation. A simple classical analysis shows that the electron orbit would decay and strike the proton in about 10^{-14} second! As we don't see atoms spontaneously collapsing in everyday life it seems that something is wrong with the classical picture. We must move on to a more advanced theory. That theory is quantum mechanics.

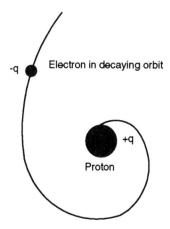

Figure 11.2   A failure of the classical physics.

11.3 The wave equation

Before studying Schrödinger's equation (the central equation of quantum mechanics as we shall study it) we need to look at the ordinary wave equation. The wave equation describes an amplitude varying in time *and* space and must therefore be a function of these two variables.

There are three very important reasons why we must fully understand the wave equation itself:

1 It is a fundamental equation of physics and as such its form must be known and understood.
2 We shall find a *solution* to the differential wave equation and from this solution we shall derive the differential wave equation *itself*! This is *exactly* the same procedure we shall use in getting to Schrödinger's equation. In the Schrödinger case we shall obtain the *fundamental* differential equation (Schrödinger equation) from a *solution* to the equation (the free particle wavefunction).
3 By studying the wave equation and by combining the results with the de Broglie findings we shall be able to construct a free particle wavefunction. Having found a free particle wavefunction we shall form the Schrödinger equation (SE).

I shall use ϕ to represent the wave disturbance (amplitude). Ψ is normally used but this could be confused with the *wavefunction* that we shall look at later. The ϕ I shall use is more general, it could describe,

(a) water waves,
(b) electromagnetic waves,
(c) waves on a string,
(d) matter waves (particles).

For case (d) ϕ becomes Ψ, the wavefunction.

Let's consider some disturbance ϕ travelling in the positive x-direction with a constant speed v. The disturbance is moving, it must be a function of position *and* time:

$$\phi = f(x, t) \tag{11.6}$$

If we take the time as zero ($t = 0$), then

$$\phi(x, t)_{(t=0)} = f(x, 0) = f(x) \tag{11.7}$$

Hence at $t = 0$ we just see the wave profile. Let's look at the travelling wave graphically and consider two frames of reference. One frame will be stationary; the other frame will be moving along with the wave. The situation is shown in Figure 11.3. The frame S' (moving frame) travels with the disturbance at speed v. In the S' frame the disturbance does not change with time, therefore $\phi \neq \phi(t)$ and we see a constant profile.

In the S' frame the coordinate is x', so

$$\phi = f(x') \text{ MOVING FRAME} \tag{11.8}$$

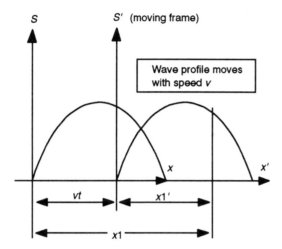

Figure 11.3   Diagram for the wave equation.

By looking at Figure 11.3, convince yourself that the relation between the moving coordinate x' and the stationary coordinate x is given by

$$x' = x - vt \qquad (11.9)$$

So ϕ can be written in terms of the stationary S frame as

$$\phi(x, t) = f(x - vt) \qquad (11.10)$$

Equation (11.10) is the most general form of the one-dimensional travelling wave.

Look at the wave. After a time δt the distance increase will be $v\delta t$ and so the function,

$$f\{(x + v\delta t) - v(t + \delta t)\} = f(x - vt)$$

remains unaltered, hence the profile remains unaltered. If the wave is travelling in the *negative* x-direction, then,

$$\phi = f(x + vt), \text{ with } v > 0 \qquad (11.11)$$

This is one of those cases where it's easy to remember the signs; they go the other way round to the direction you're considering!

We have in equation (11.10) the most general form of the travelling wave. From this functional form we are able to derive the differential form of the wave equation. We do this by differentiating the functional form twice with respect to time and twice with respect to x. The necessary maths is given below:

Using $x' = x \mp vt$, then

$$\frac{\partial x'}{\partial x} = 1, \quad \frac{\partial x'}{\partial t} = \mp v$$

Differentiating with respect to x:

$$\frac{\partial \phi}{\partial x} = \frac{\partial f}{\partial x'} \cdot \frac{\partial x'}{\partial x} = \frac{\partial f}{\partial x'}$$

and differentiating again with respect to x gives

$$\frac{\partial^2 \phi}{\partial x^2} = \frac{\partial^2 f}{\partial x'^2}$$

Differentiating with respect to time gives

$$\frac{\partial \phi}{\partial t} = \frac{\partial f}{\partial x'} \cdot \frac{\partial x'}{\partial t} = \mp v \cdot \frac{\partial f}{\partial x'}$$

and differentiating once again with respect to time yields

$$\frac{\partial^2 \phi}{\partial t^2} = v^2 \cdot \frac{\partial^2 f}{\partial x'^2}$$

Combining these two results gives us what we're after, the one-dimensional differential wave equation:

$$\frac{\partial^2 \phi}{\partial x^2} = \frac{1}{v^2} \cdot \frac{\partial^2 \phi}{\partial t^2} \qquad (11.12)$$

It's simple to remember where v goes, since $x = vt$ the v and the t go together (look at the form of the equation again).

It's very straightforward to extend the one-dimensional equation to three dimensions:

$$\phi = \phi(x, y, z, t)$$

$$\frac{\partial^2 \phi}{\partial x^2} + \frac{\partial^2 \phi}{\partial y^2} + \frac{\partial^2 \phi}{\partial z^2} = \frac{1}{v^2} \cdot \frac{\partial^2 \phi}{\partial t^2}$$

$$\left(\frac{\partial^2}{\partial x^2} + \frac{\partial^2}{\partial y^2} + \frac{\partial^2}{\partial z^2} \right) \phi = \frac{1}{v^2} \cdot \frac{\partial^2 \phi}{\partial t^2}$$

The above equation can be shortened by using the Laplacian operator in Cartesian coordinates, where

$$\nabla^2 = \frac{\partial^2}{\partial x^2} + \frac{\partial^2}{\partial y^2} + \frac{\partial^2}{\partial z^2}$$

Using the Laplacian operator, the three-dimensional wave equation becomes

$$\nabla^2 \phi = \frac{1}{v^2} \cdot \frac{\partial^2 \phi}{\partial t^2} \qquad (11.13)$$

11.4 Harmonic waves

Since we know the most general functional form for solutions of the wave equation, we can substitute other well-known functions for f in $f(x - vt)$. What functions could we substitute, and why should we? The sine and cosine functions are obvious candidates for substitution and they are particularly important because any periodic waveform (square, triangular, rectangular, etc.) can be made up from a series of sine and cosine functions. Such a series is called a *Fourier series*. Wherever I write sin in this section we could equally well write cos, so for a general form of a harmonic travelling wave we could have

$$\phi = A \sin\{k(x \mp vt)\} \tag{11.14}$$

Here A is a constant, and k is called the propagation constant, where

$$k = \frac{2\pi}{\lambda} \tag{11.15}$$

Calculation

Prove the above relation for the propagation constant.

The repetitive spatial unit of a harmonic wave is its wavelength λ. Increasing all x by λ will reproduce the same wave. In exactly the same way, increasing the argument of the sine function by 2π will reproduce the wave. We can write this mathematically as

$$A \sin k\{(x + \lambda) + vt\} = A \sin\{k(x + vt) + 2\pi\}$$

Expanding the equation gives

$$A \sin(kx + ky + kvt) = A \sin(kx + kvt + 2\pi)$$

Hence,

$$k\lambda = 2\pi$$

Or, in terms of the propagation constant,

$$k = \frac{2\pi}{\lambda}$$

Alternatively, the waveform is periodic in time with period T. Increasing all t by T would also reproduce the same waveform. We can go through the same mathematics as above to show that

$$v = \nu\lambda \tag{11.16}$$

where the frequency $\nu = 1/T$.

Other parameters are often used, for example

$$\omega = \text{angular frequency} = 2\pi\nu$$

$$\varkappa = \text{wave number} = \frac{1}{\lambda}$$

The following forms are therefore all equivalent,

$$\phi = A \sin\{k(x \mp vt)\}$$

$$\phi = A \sin\left\{2\pi\left(\frac{x}{\lambda} \mp \frac{t}{T}\right)\right\} \tag{11.17}$$

$$\phi = A \sin(kx \mp \omega t) \tag{11.18}$$

As stated earlier, these could all be cos functions rather than sin functions; it's merely a shift of $\pi/2$ radians.

11.5 Complex representation

The previous section showed that equations such as

$$\phi = A \cos(kx - \omega t)$$

$$\phi = A \sin(kx - \omega t)$$

represent travelling harmonic waves. There is a relation between harmonic functions and exponential functions containing the imaginary number, i, given by Euler's formula. Euler's formula states,

$$\exp(i\theta) = \cos\theta + i \sin\theta \tag{11.19}$$

So we could write down the equations for our harmonic travelling waves in exponential form as

$$\phi = A \exp i(kx - \omega t) \tag{11.20}$$

which can be written using Euler's formula as

$$\phi = A \cos(kx - \omega t) + iA \sin(kx - \omega t) \tag{11.21}$$

The real part of ϕ is given by

$$\Re\{\phi\} = A \cos(kx - \omega t) \tag{11.22}$$

and the imaginary part by

$$\Im\{\phi\} = A \sin(kx - \omega t) \tag{11.23}$$

The reason for going through all this is so that we can now work with the exponential form, which is a lot easier than manipulating harmonic functions. By working with exponentials we can extract either the sine terms or cosine terms at any point in a calculation if we want to by either looking at the imaginary or real parts of the equation.

There is a little more to this exercise than just looking at different forms of waves, however! The exponential form we have just produced above is going to provide us with a free particle (not acted on by any external forces) wavefunction. The free particle wavefunction is itself the solution to an all embracing powerful equation which is the framework of quantum mechanics, the Schrödinger equation. First rewrite the above equation in a slightly different form:

$$\phi = A \exp -i\omega(t - x/v) \tag{11.24}$$

Replace ω by $2\pi\nu$ and ν by $\lambda\nu$.

$$\phi = A \exp{-2\pi i(\nu t - x/\lambda)} \tag{11.25}$$

We are now in a position to turn the above equation into a free particle wavefunction! We know what ν and λ are in terms of the total energy E and momentum p *of a particle*:

$$E = h\nu$$

$$\lambda = \frac{h}{p}$$

Substitute the above into the exponential wave equation and change the ϕ to a Ψ since we are now forming a free particle wavefunction.

$$\Psi = A \exp{-\frac{i}{\hbar}(Et - px)} \tag{11.26}$$

This equation is the mathematical description of the wave equivalent of an unrestricted *particle* of total energy E and momentum p moving in the positive x-direction. It is the *wavefunction* for a free particle moving in the positive x-direction and we shall use this free particle wavefunction as a starting point in constructing Schrödinger's equation.

Recap briefly what we did to construct the free particle wavefunction. We started with the complex form of the ordinary wave equation. We did this because it is the most convenient form to work on. The equation we chose was for an unrestricted wave travelling in the positive x-direction. We then used the results of de Broglie to change parameters usually associated with *waves* (such as wavelength) to parameters usually associated with *particles* (such as momentum). The resulting equation was called a free particle wavefunction, an equation describing an unrestricted particle moving in the positive x-direction. This all seems fairly reasonable and logical, which in fact it is. It is hardly rigorous, however, but this approach is merely to show you the overall machinery, not the inner workings in detail.

11.6 Schrödinger's equation

We have looked in some depth at the wave equation and some of its solutions. From one of these solutions we have constructed the *wavefunction* for a totally free particle of total energy E and momentum p moving in the $+x$-direction.

In practice we aren't particularly interested in totally free particles; we are more interested in particles under the influence of forces such as an electron bound to an atomic nucleus. We therefore need to introduce the particle *potential energy* into the free particle wavefunction. How can we do this?

At speeds small compared to the speed of light (in other words we are going to do non-relativistic quantum mechanics), the *total* energy of a particle is the sum of its kinetic energy $p^2/2m$ and its potential energy V, where V is in general a function of position x and time t.

$$E = \frac{p^2}{2m} + V \quad \text{TOTAL ENERGY} \tag{11.27}$$

Differentiate the free particle wavefunction, twice with respect to x, once with respect to t (and look back at how we arrived at the ordinary wave equation). Performing the differentiations gives

$$\frac{\partial \Psi}{\partial x} = \frac{ip}{\hbar} \Psi \tag{11.28}$$

$$\frac{\partial^2 \Psi}{\partial x^2} = -\frac{p^2}{\hbar^2} \Psi \tag{11.29}$$

$$\frac{\partial \Psi}{\partial t} = -\frac{iE}{\hbar} \Psi \tag{11.30}$$

Multiply both sides of the total energy equation (11.27) by Ψ:

$$E\Psi = \frac{p^2 \Psi}{2m} + V\Psi \tag{11.31}$$

and substitute the partial differential forms for E and p as found above to obtain

$$\frac{\hbar}{i} \cdot \left(\frac{\partial \Psi}{\partial t}\right) = \frac{\hbar^2}{2m} \cdot \left(\frac{\partial^2 \Psi}{\partial x^2}\right) - V\Psi \tag{11.32}$$

At last! This is the equaton we have been working towards. This is the time-dependent Schrödinger equation in one dimension. Using this equation we can probe the secrets of the atom and understand how particles can 'tunnel' or have specific (quantized) energy levels. Just before we start using this equation we need to see some other (simpler) forms and how we actually operate on the equation.

In three dimensions (recall the earlier work on the ordinary wave equation) the Schrödinger equation becomes

$$\frac{\hbar}{i} \cdot \left(\frac{\partial \Psi}{\partial t}\right) = \frac{\hbar^2}{2m} \cdot \left(\frac{\partial^2 \Psi}{\partial x^2} + \frac{\partial^2 \Psi}{\partial y^2} + \frac{\partial^2 \Psi}{\partial z^2}\right) - V\Psi \tag{11.33}$$

This can be simplified, as we saw before, using the Laplacian operator:

$$\frac{\hbar}{i} \cdot \left(\frac{\partial \Psi}{\partial t}\right) = \left(\frac{\hbar^2}{2m}\right) \nabla^2 \Psi - V\Psi \tag{11.34}$$

which gives us the time-dependent Schrödinger equation in three dimensions.

Notice at this point that the *form* of the Schrödinger equation is *not* the same as the *form* for the ordinary wave equation. The Schrödinger equation is only first order in t! The Schrödinger equation may have wave-like solutions but it is *not* a wave equation. Recall that

$$\nabla^2 \phi = \frac{1}{c^2} \cdot \frac{\partial^2 \phi}{\partial t^2}$$

is the form of the wave equation. The Schrödinger equation has the form of the heat conduction or diffusion equation:

$$\nabla^2 u = \frac{1}{k} \cdot \frac{\partial u}{\partial t}$$ (11.35)

11.7 Steady-state form of the Schrödinger equation

In a lot of physical situations, and in the cases we shall study later, the energy of a particle may not vary with time, although it may vary with position. If this is the case then the Schrödinger equation can be considerably simplified by removing all reference to the time t.

Returning to the one-dimensional *wavefunction* Ψ of an unrestricted particle:

$$\Psi = A \exp{-\frac{2\pi i}{h}(Et - px)}$$

which can be expanded to separate the space and time components as below:

$$\Psi = A \exp{-\left(\frac{iE}{\hbar}\right)} t \cdot \exp{\left(\frac{ip}{\hbar}\right)} x$$ (11.36)

This in turn can be written as

$$\Psi = \psi \cdot \exp{-\left(\frac{iE}{\hbar}\right)} t$$ (11.37)

where Ψ is seen to be the product of a time-dependent function and a position-dependent function ψ. Substitute this Ψ into the time-dependent Schrödinger equation to obtain

$$-E\psi \cdot \exp{-\left(\frac{iE}{\hbar}\right)} \cdot t = \frac{\hbar^2}{2m} \cdot \left(\frac{\partial^2 \psi}{\partial x^2}\right) \exp{-\left(\frac{iE}{\hbar}\right)} \cdot t - V \cdot \psi \exp{-\left(\frac{iE}{\hbar}\right)} \cdot t$$ (11.38)

The time function cancels throughout and a slight rearrangement gives

$$\frac{\partial^2 \psi}{\partial x^2} + \frac{2m}{\hbar^2}(E - V)\psi = 0$$ (11.39)

which is the steady-state Schrödinger equation in one dimension. In exactly the same way as before, the extension to three dimensions is straightforward and leads to

$$\nabla^2 \psi + \frac{2m}{\hbar^2}(E - V)\psi = 0$$ (11.40)

11.8 The wavefunction

We have now arrived at the point where it is not prudent to think too deeply about the physics, but instead to simply use the tools provided. Only one quantity needs description and that is the particle *wavefunction*. Schrödinger's equation (SE) is the fundamental differential equation of quantum mechanics and it is a differential equation involving the particle wavefunction.

The wavefunction of a particle is denoted by Ψ, which in itself has no physical meaning (I know this is already beginning to sound pretty bad, but bear with it), but its absolute magnitude squared $|\Psi|^2$ or its complex square $(\Psi^*\Psi)$ gives us the probability of finding the particle at the point being considered at the time being considered. Let us look at a simple example. Suppose we know that a particle, described by the wavefunction Ψ, definitely exists somewhere in space. Then the following equation must be true:

$$\int_{-\infty}^{\infty} |\Psi|^2 dV = 1 \tag{11.41}$$

where we have integrated the absolute square of the wavefunction (giving us the probability of finding the particle) over all space (dV = all space). Since the particle exists the integral equals unity. We need not have integrated over all space; we could have integrated over a small volume in space and hence found the probability of the particle existing in the small volume.

The final hurdle now is to use Schrödinger's equation in *real* physical situations and find solutions to the equation. By looking at the solutions we can obtain all the information we need about the particle being described.

The way to use quantum mechanics to solve physical problems is as follows:

1 Write down the appropriate Schrödinger equation. There are the time-dependent and time-independent forms.
2 Apply the physical boundary conditions to obtain the solutions to Schrödinger's equation under the given circumstances.
3 These solutions are the particle wavefunctions. They tell us all we need to know about the particle under the given conditions.

We shall now use the Schrödinger equation in the one-dimensional, steady-state form as found above in order to solve the particle-in-a-box problem.

11.9 The particle-in-a-box

Figure 11.4 shows a one-dimensional potential well containing a particle that is called the particle-in-a-box. The walls of the well are infinitely high meaning that the particle is effectively constrained within the potential walls and cannot penetrate them (this is not true if the walls are not infinitely high, as we shall see). We may take the particle potential energy to be any convenient value in this situation and so it is convenient to choose $V = 0$ between the potential walls.

Putting $V = 0$ in the steady-state Schrödinger equation yields

$$\frac{d^2\psi}{dx^2} + \frac{2m}{\hbar^2} E\psi = 0 \tag{11.42}$$

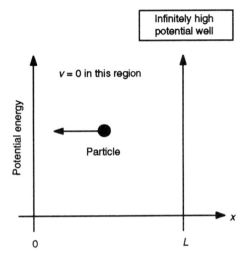

Figure 11.4 Particle-in-a-box.

where we have now used a total differential rather than a partial differential because ψ is a function of x only in this case. There are two possible solutions for ψ in equation (11.42); these are

$$\psi = A \sin\sqrt{\frac{2mE}{\hbar^2}} \cdot x \qquad (11.43)$$

and

$$\psi = B \cos\sqrt{\frac{2mE}{\hbar^2}} \cdot x \qquad (11.44)$$

Their sum is also a solution. These solutions can be verified by back substitution into (11.42) but equation (11.42) and its solutions should already be well-known to you since the form is the same as for the *simple harmonic oscillator*!

   Working our way down the checklist on solving physical problems using quantum mechanics we see that the next thing to do is to apply the physical boundary conditions. Here it is appropriate to mention that the correct application of the boundary conditions is perhaps the most difficult part of the solving real physical problems! However, in this case the boundary conditions are clear. The particle cannot exist at, or beyond the potential barriers. If the particle cannot exist beyond these limits, then the wavefunction (which describes the particle) cannot exist either. Therefore, for $x = 0$ and for $x = L$, where L is the width of the potential well, $\psi = 0$. Since $\psi = 0$ at $x = 0$ we know we must take the sine solution (equation (11.43)) because the cosine solution is non-zero at $x = 0$.

   So we now have for our wavefunction equation (11.43)

$$\psi = A \sin\sqrt{\frac{2mE}{\hbar^2}} \cdot x$$

and at this stage we apply the second boundary condition, $\psi = 0$ at $x = L$. When do sine arguments become zero? The answer is when they are equal to integer multiples of π. Hence ψ will be zero at $x = L$ when

$$\sqrt{\frac{2mE}{\hbar^2}} \cdot L = \pi, \, 2\pi, \, 3\pi, \, \ldots \qquad (11.45)$$

This result immediately gives us something rather important. Equation (11.45) gives us the energy eigenvalues (the allowed energy-levels) for the particle-in-a-box! Rewrite equation (11.45) in terms of the energy:

$$E_n = \frac{n^2 h^2}{8mL^2} \text{ where } n = 1, 2, 3, \ldots \qquad (11.46)$$

Each n gives rise to an allowed energy level for the particle-in-a-box. The integer n is called the particle's principal quantum number. Note that by merely constraining a particle's motion we have restricted its allowed energies to those given by equation (11.46). A particle restricted in this way cannot have arbitrary energy! Think now of an electron restricted in its motion about an atomic nucleus. It is not now too difficult to see why the electron can only have certain well-defined energy-levels in its motion about the atom. Some very simple quantum mechanics is quickly giving us some very deep physical insights, as promised.

There is another important result from equation (11.46). Note that $n = 0$ is not allowed! If $n = 0$ then the wave function ψ would have to be zero and the particle would not exist. This excludes the possibility of $E = 0$ as well. By merely constraining a particle you ensure that its energy can never be zero! Does this phenomenon have any bearing in reality? Yes it does! As temperatures are taken down towards absolute zero atomic electrons do not collapse on to the nucleus; even at zero Kelvin they have some energy left, the *zero point energy*, a manifestation of the fact that $E = 0$ is not allowed, even at absolute zero, for a particle in a potential well.

The above results fly in the face of classical theory, where in a similar physical situation the particle could have any energy *including zero*. This is a suitable place to introduce another quantum mechanical result that you have heard of before, the Heisenberg uncertainty principle. One form of the principle is given by

$$\Delta x \Delta p_x \geq \hbar \qquad (11.47)$$

which in words says that the product of the uncertainty in the position of a particle and its momentum in the same direction cannot be less than Planck's constant divided by 2π. Our particle-in-a-box is uncertain in position to the width of the box, that is, L. Hence, the uncertainty in its momentum must be

$$\Delta p_x \geq \frac{\hbar}{L} \qquad (11.48)$$

and, since L is finite, the particle momentum must be finite and again the particle cannot have zero energy. We have used just a tiny bit of the quantum mechanics, yet look at these results! It's really difficult (as I warned you) not to start getting carried away now. If the Universe is a closed system, then all particles are constrained in a (gravitational) potential well of the size of the Universe, and so by the Heisenberg uncertainty principle no particle in existence can have zero energy (though to be honest the minimum allowed energy value will be extremely small). Let's return to reality and continue our investigation of the particle-in-a-box where we left off, having just found the allowed energy-levels. By just restricting our particle's movement we found this set of allowed energies. The particle can only have one of these allowed energies and no other. At this stage you should now be trying to think of situations to 'beat' the theory. Never take anything at face value, always try to disprove it; if you can't, you've got a good theory. An obvious test of quantum mechanics is to bring the equations from the microscopic world up into the macroscopic world and see if they hold up! How can we do this for the particle-in-a-box? Well, let's scale up the particle for a start and let's consider a marble weighing 30 g instead. And how about the potential barriers? Let's have the marble running along a pipe with the ends blocked off, the pipe length being 30 cm. Make one. Shake the marble about. Can you 'feel' any quantized energy-levels? Is the zero of energy not allowed; in other words is the marble always moving? Doesn't seem like much more than you would expect is happening. The marble runs up and down the pipe as you tilt it and that's about all. If that's the case then the quantum equations had better agree or we'll be in trouble!

We have a 30 g marble in a 30 cm 'box'. Substitute values into equation (11.46) to find the allowed energy-levels. We get

$$2.03 \times 10^{-65} \times n^2 \text{ joules} \tag{11.49}$$

So the minimum marble energy is for the $n = 1$ case, i.e.

$$2.03 \times 10^{-65} \text{ joules} \tag{11.50}$$

With this minute energy the marble velocity would be $1.16 \times 10^{-32} \text{ ms}^{-1}$, which is of course indistinguishable from stationary! OK that didn't work, so what about the energy quantization? If we pick a realistic marble speed of 30 cm s^{-1} the corresponding energy is 1.35×10^{-3} joules and the associated quantum number from equation (12.8) is $n = 10^{29}$! At such very large quantum numbers the energy-levels are so close together that the discrete nature is lost and they become a continuum (refer back to the case of what the energy-levels did when we brought large numbers of atoms close together).

This shows why quantum effects are not noticeable as everyday occurrences and this is why the simpler Newtonian mechanics is so successful.

Let's go back to that very large quantum number we found in looking at the marble in a box. Such large quantum numbers usually appear when we apply quantum mechanics to macroscopic problems and the results we obtain, as we have just seen, are in accordance with observation. This is in fact a powerful quantum mechanical principle, that in the limit of large quantum numbers things should appear to us as they normally do in the macroscopic world. In quantum mechanics this is called the *correspondence principle*.

After this brief excursion looking in a little detail at the energy eigenvalues let's return to the particle-in-a-box equations. We found that the allowed wavefunctions were given by (11.43):

$$\psi = A \sin\sqrt{\frac{2mE}{\hbar^2}} \cdot x$$

with the energy eigenfunctions represented by equation (11.46):

$$E_n = \frac{n^2 h^2}{8mL^2}$$

If we now substitute E_n for E in equation (11.43) we obtain the eigenfunctions or the allowed wavefunctions for the particle-in-a-box. These functions when squared and plotted against the variable x will show us the probability of finding the particle between the barrier walls as a function of x. First make the substitution to obtain

$$\psi_n = A \sin\frac{n\pi x}{L} \quad \text{EIGENFUNCTIONS} \qquad (11.51)$$

Finally, if we are to plot these functions out we have to get rid of the constant A. The process we shall use is called *normalization*. We know, since the particle exists, that if we integrate the absolute magnitude of the wavefunction squared, over all x, we shall get unity. Performing the integration we shall also get an equation containing A and can therefore substitute for A, as follows.

First integrate the absolute magnitude squared of the wavefunction over all x:

$$\int_{-\infty}^{\infty} |\psi_n|^2 \, dx = \int_0^L |\psi_n|^2 \, dx \qquad (11.52)$$

The range of integration has been changed from infinity to the length of the well since the particle only exists between the well walls. Now substitute for the eigenfunctions (11.51) and perform the integration:

$$A^2 \int_0^L \sin^2\left(\frac{n\pi x}{L}\right) dx = \frac{A^2 L}{2} \qquad (11.53)$$

We also know that this integral equals unity, hence

$$A = \sqrt{\frac{2}{L}} \qquad (11.54)$$

And so finally, substituting for A, we obtain the *normalized wavefunctions* for the particle-in-a-box:

$$\psi_n = \sqrt{\frac{2}{L}} \sin\left(\frac{n\pi x}{L}\right) \qquad (11.55)$$

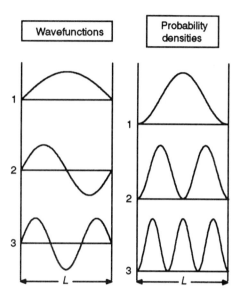

Figure 11.5 Particle-in-a-box with infinite barriers.

We can now plot out these wavefunctions, or, more importantly the *square* of these wavefunctions, which gives us probability values. Figure 11.5 is such a plot. Notice that a particle in the lowest energy level ($n = 1$) is most likely to be found in the middle of the box, while a particle with $n = 2$ is never there!

If the height of the potential barriers is not infinite, there is the possibility of barrier penetration, i.e. there is the finite probability of finding the particle beyond the barrier. This situation is shown in Figure 11.6. The thing to note is the rapid (exponential) drop in probability in finding the particle beyond the barrier. This means that for high, wide, barriers the particle will not get through because of this exponential drop in probability. However, if the barrier is not too wide, or high, for example the potential barrier that confines the nucleus of uranium atoms, then occasionally barrier penetration *can* take place, and an alpha particle is emitted from the uranium nucleus!

What does all this have to do with electronic devices? Today there are some very powerful techniques for growing semiconductor layers such as molecular beam epitaxy (MBE) and metalorganic chemical vapour deposition (MOCVD). The semiconductor layers can be grown so thin that the electrons (or holes) within them behave like particles-in-a-box and show energy quantization. These effects are utilized in quantum-well lasers and certain types of detectors. We shall take a brief look at the potential of quantum-well lasers.

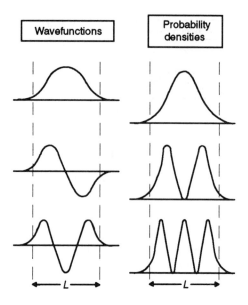

Figure 11.6 Particle-in-a-box with finite barriers.

11.10   The quantum-well laser

A simplified band diagram for a quantum-well laser is shown in Figure 11.7. Although it may look a little complicated you should already be able to construct this sort of diagram from our earlier work looking at the semiconductor laser band diagrams. Aluminium gallium arsenide (AlGaAs) has a bigger band gap than straight gallium arsenide (GaAs), hence a potential well for electrons appears in the conduction band and a potential well for holes appears in the valence band, as shown in the diagram. So far this is not very much different from the ordinary heterostructure laser covered in Chapter 10. In fact, the only difference is the width of the GaAs active region, in this case we make the width L of the active region of the order 50–100 Angstrom. We are able to make such thin layers using modern epitaxial growth techniques, as already mentioned, and in doing so we are able to produce particle-in-a-box-type energy quantization in the valence and conduction bands. How does this help us in any way? Well, for a normal *heterostructure laser* we are stuck with laser light energy equal to the *band gap of the active layer*. Now look at the quantum-well laser. The energy quantization has produced some allowed discrete quantum states in the conduction and valence bands. What's even more important is that from our particle-in-a-box studies we know that electrons will want to be in the $n = 1$ state (or higher, since $n = 0$ is not allowed), and similarly for the holes. The electron–hole recombination which is the most favoured occurs between the $n = 1$

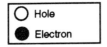

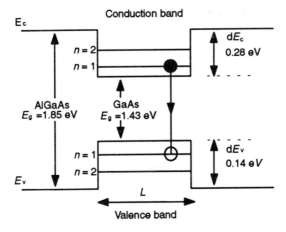

Figure 11.7   Quantum-well laser.

level in the conduction band, and the $n = 1$ level in the valence band. Therefore, the laser light output has an energy *greater* than the GaAs band gap! We are able to *tune* the laser output wavelength (within limits) merely by altering the *width* of the active layer!! No wonder there is so much interest in these new devices; we aren't stuck with wavelengths given to us by Nature, we are able to tune them. What are the tuning limits? We can crudely estimate limits in the following way. If the well-width is so narrow that the $n = 1$ level is within kT of the top of the well then we lose particle confinement in the active layer, and hence we lose the heterojunction action. If the valence band discontinuity is of order $0.14\,\text{eV}$, then the conduction band discontinuity will be about $0.28\,\text{eV}$. With your particle-in-a-box knowledge you should now be able to calculate how thin you can make the active region before losing heterojunction action. Happy calculating.

Derivation of electron and hole carrier densities

We want to be able to calculate how many electrons there are in the conduction band (or how many holes there are in the valence band) of a semiconductor. The integral we're going to have to work out is

$$n = \int_0^{E_{\text{TOP}}} N(E) P_{\text{fd}}(E) \, dE \tag{A1.1}$$

where

the integral is from the bottom ($E_C = 0$, band edge) of the conduction band to the *top* of the conduction band,
n = the electron density in the conduction band (cm^{-3}),
$N(E)$ = the density of allowed states per unit volume (cm^{-3}),
$P_{\text{fd}}(E)$ = the Fermi–Dirac function.

We don't know what $N(E)$, the density of allowed states (per unit volume), is. We'll have to calculate it.

In order to calculate n we use the concept of momentum space, see Figure A1.1. DON'T PANIC!! This isn't very different from ordinary x, y, z space you're used to dealing with. The difference is that we're only dealing with momenta (not distances so the axes are labelled p_x, p_y, p_z. Consider just the x-direction in a semiconductor crystal for a moment.

The electron in a crystal may be described by standing waves (quantum mechanics), and if the crystal length is L with an electron wavelength of λ, we have, for standing waves in the x-direction,

$$\frac{L}{\lambda} = n_x, \; n_x = \text{INTEGER} \tag{A1.2}$$

The electron wavelength is related to the electron momentum by

$$\lambda = \frac{h}{p_x}, \; p_x = \text{momentum in } x\text{-direction} \tag{A1.3}$$

Substitute for λ in (A1.2),

$$L p_x = h n_x \tag{A1.4}$$

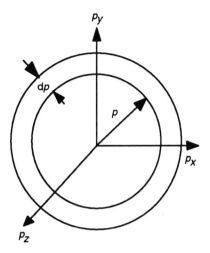

Figure A1.1 Momentum space.

The incremental *momentum*, dp_x, required for unity increase in n_x is

$$L \, dp_x = h \tag{A1.5}$$

So, for the three-dimensional cube of semiconductor with side L:

$$L^3 \, dp_x \, dp_y \, dp_z = h^3 \tag{A1.6}$$

So the incremental *volume* in momentum space for a unit cube ($L = 1$) is

$$dp_x \, dp_y \, dp_z = h^3, \text{ UNIT VOLUME} \tag{A1.7}$$

Return to Figure A1.1. The incremental volume of momentum space, dV, contained between the spheres of radius p and $p + dp$ is

$$dV = 4\pi p^2 \, dp \tag{A1.8}$$

The number of *allowed* energy states contained in this volume is $N(E)$:

$$N(E) = \frac{2 \times (4\pi p^2 \, dp)}{h^3} \tag{A1.9}$$

where

the 2 comes from two allowed energy states per electron,
$h^3 = $ volume per state,
$4\pi p^2 \, dp = $ incremental volume.

Substitute for p using $E = p^2/2m_n$, where m_n is the electron effective mass:

$$N(E) = 4\pi \left(\frac{2m_n}{h^3}\right)^{\frac{3}{2}} E^{\frac{1}{2}} \tag{A1.10}$$

where $N(E)$ is the density of allowed energy states per unit volume. Notice that the density of states is parabolic ($E^{0.5}$) with respect to energy, so there are more allowed states per unit volume as you move up the band (in energy). For incremental energy dE:

$$N(E)\,dE = 4\pi\left(\frac{2m_n}{h^2}\right)^{\frac{3}{2}} E^{\frac{1}{2}}\,dE \tag{A1.11}$$

We're nearly ready to carry out integration (A1.1) but first we need two changes:

1 Because P_{fd} *decreases* exponentially with *increasing* E, and since most electrons are near the bottom of the conduction band, we can replace E_{TOP} by infinity without seriously affecting the result.
2 We are forced to look at material that is *not* highly doped so that $E - E_F > 3kT$. This allows us to replace P_{fd} by P_{mb} (Maxwell–Boltzmann function), which means the integral will now have an analytical solution.

Let's finally get on with the integration:

$$n = \int_0^\infty N(E)P_{mb}(E)\,dE \tag{A1.12}$$

Substituting

$$n = 4\pi\left(\frac{2m_n}{h^2}\right)^{\frac{3}{2}}\int_0^\infty E^{\frac{1}{2}}\exp-\left(\frac{E - E_F}{kT}\right)dE \tag{A1.13}$$

Let $x = E/kT$:

$$n = 4\pi\left(\frac{2m_n}{h^2}\right)^{\frac{3}{2}}(kT)^{\frac{3}{2}}\exp\left(\frac{E_F}{kT}\right)\int_0^\infty x^{\frac{1}{2}}\exp-(x)\,dx \tag{A1.14}$$

The integral is a standard:

$$\int_0^\infty x^{\frac{1}{2}}\exp-(x)\,dx = \frac{\pi^{\frac{1}{2}}}{2} \tag{A1.15}$$

So, finally

$$n = 2\left(\frac{2\pi m_n kT}{h^2}\right)^{\frac{3}{2}}\exp\left(\frac{E_F}{kT}\right) \tag{A1.16}$$

We can refer to the bottom of the conduction band as E_C instead of $E = 0$, in which case we get

$$n = 2\left(\frac{2\pi m_n kT}{h^2}\right)^{\frac{3}{2}}\exp-\left(\frac{E_C - E_F}{kT}\right) \tag{A1.17}$$

The quantity $2(2\pi m_n kT/h^2)^{\frac{3}{2}}$ is called the *effective density of states* in the conduction band, N_C. Note that this *is not* the *actual* density of states we found earlier. Hence,

$$n = N_C\exp-\left(\frac{E_C - E_F}{kT}\right) \tag{A1.18}$$

A similar calculation gives us the hole density p in the valence band. If the hole effective mass is m_n, then

$$p = 2\left(\frac{2\pi m_p kT}{h^2}\right)^{\frac{3}{2}} \exp -\left(\frac{E_F - E_V}{kT}\right) \tag{A1.19}$$

$$N_V = 2\left(\frac{2\pi m_p kT}{h^2}\right)^{\frac{3}{2}} \tag{A1.20}$$

$$p = N_V \exp -\left(\frac{E_F - E_V}{kT}\right) \tag{A1.21}$$

where N_V is the *effective density of states* in the valence band.

The Einstein radiative transition coefficients

We'll start by looking at Figure A2.1, which is a simplified version of Figure 10.1. Figure A2.1 shows two energy-levels E_1 and E_2, which are *radiatively connected*. By this I mean that electron transitions between these two levels are accompanied either by the absorption or emission of photons. The photon energy of these transitions is simply

$$E_{\text{PHOTON}} = E_2 - E_1 = h\nu_{21} \tag{A2.1}$$

where ν_{21} is the frequency of the photon emitted when an electron drops from level 2 to level 1.

Just as we did in Chapter 10, we'll consider the three processes of *absorption, stimulated emission and spontaneous emission*. We'll also consider the two energy-level system of Figure A2.1 to be in *thermal equilibrium*. I am doing this so that I can use thermal equilibrium equations, specifically the *black-body radiation equation*, which gives the energy density of photons emitted by a black-body (a body in thermal equilibrium) at temperature T. What we're going to do (in words) is this. By considering the system of Figure A2.1, and how it interacts with equilibrium (black-body) radiation, we're going to derive an equation for the incoming radiation energy density in terms of the A and B coefficients shown in the diagram. We will then be able to compare this equation with the black-body radiation equation (which we know), and will then be able to see what the A and B coefficients imply.

Let's recall what the black-body radiation equation actually looks like:

$$\rho(\nu)\, d\nu = \frac{(8\pi n^3 h \nu^3)}{c^3} \frac{1}{e^{h\nu/kT} - 1}\, d\nu \tag{A2.2}$$

where

$\rho(\nu)\, d\nu$ is the energy density of the equilibrium photons (joules per cubic metre) of frequency ν in the frequency interval $d\nu$.

c is the velocity of light *in-vacuo*.

n is the material's refractive index.

This *energy density* is therefore simply related to the *number density* (via the

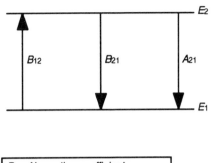

B_{12} Absorption coefficient
B_{21} Stimulated emission
A_{21} Spontaneous emmision

Figure A2.1   Einstein coefficients.

photon energy) of the incoming photons at frequency v. Now the *actual* photon frequency we're going to be interested in is v_{21}, the frequency that will cause optical transitions between levels 1 and 2. I'll just call this frequency v in the following derivation.

Considering the *absorption* process first, the *rate* at which electrons will be transferred from level 1 to level 2 will depend on the following:

(a) the number of incident *effective* photons,
(b) the number of electrons in level 1.

If the number of electrons in level 1 is N_1, then the rate will be *proportional* to the product of (a) and (b). Hence the rate will be *equal* to

$$ABSORPTION\ RATE\ =\ N_1 \rho(v) B_{12} \qquad (A2.3)$$

where B_{12} is the *absorption rate coefficient* for the process.

Now consider the *stimulated emission* process. The rate at which electrons will be transferred from level 2 to level 1 via this process will depend on the following:

(c) the number of incident *effective* photons,
(d) the number of electrons in level 2.

If the number of electrons in level 2 is N_2, then the rate will be proportional to the product of (c) and (d). Hence the rate will be equal to

$$STIMULATED\ EMISSION\ RATE\ =\ N_2 \rho(v) B_{21} \qquad (A2.4)$$

where B_{21} is the *stimulated emission rate coefficient* for the process.

Finally, let's consider the *spontaneous emission* process. We know from Chapter 10 that this process differs from the other two in that no incident photon is required for the electron to fall from level 2 to level 1. This makes the form of the spontaneous emission rate particularly simple:

$$SPONTANEOUS\ EMISSION\ RATE\ =\ N_2 A_{21} \qquad (A2.5)$$

Since our two level system is in thermal equilibrium, the rate of transitions from level 1 to level 2 must equal the transition rate from level 2 to level 1:

$$N_1\rho(v)B_{12} = N_2\rho(v)B_{21} + N_2A_{21} \tag{A2.6}$$

Making $\rho(v)$ the subject of the equation gives

$$\rho(v) = \frac{A_{21}/B_{21}}{(B_{12}N_1/B_{21}N_2) - 1} \tag{A2.7}$$

We can immediatly substitute for N_1/N_2 using the Boltzmann relation

$$\frac{N_1}{N_2} = \exp\frac{(E_2 - E_1)}{kT} \tag{A2.8}$$

Remembering that $E_2 - E_1$ is just hv, then substitute (A2.8) into (A2.7):

$$\rho(v) = \frac{A_{21}/B_{21}}{B_{12}/B_{21}\left[\exp\left(\frac{hv}{kT}\right) - 1\right]} \tag{A2.9}$$

We can now compare equation (A2.9) with equation (A2.2) to find

$$B_{12} = B_{21} \tag{A2.10}$$

and,

$$A_{21} = B_{21}\frac{8\pi hn^3v^3}{c^3} \tag{A2.11}$$

Equation (A2.10) tells us what we wanted to know from Chapter 10, namely that the absorption and spontaneous emission rates are equal for a system in thermal equilibrium.

Equations (A2.10) and (A2.11) were first given by Einstein (1917).

The four-level laser system

The object of this appendix is to see why it is easier to achieve population inversion in a four-level system than a three-level system. Figure A3.1 shows the energy-level configuration for a four-level laser system. If you refer back to Figure 10.4 you'll see that many things are similar, *with the exception of the lower laser-level*. Notice that in this four-level system, the lower laser-level is no longer the ground-state, it lies some way above the ground-state-level. In fact, if the lower laser-level is several kT *above* the ground-state-level, then in thermal equilibrium *the lower laser-level is empty*. If we now pump this four-level system (in exactly the same way as a three-level system) and if the level 4 to ground-state transition

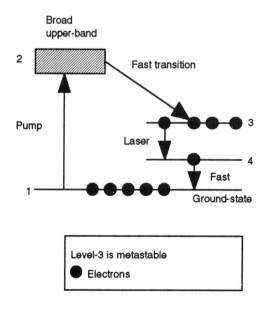

Figure A3.1 Four-level laser system.

is fast, then level 4 effectively remains empty. This means that as soon as we get *any* electrons in level 3 *we have a population inversion with respect to level 4* (the lower laser-level). As soon as we get the threshold density of electrons in level 3, therefore, we get laser action.

Recall that we needed to get *more* than half the ground-state density of electrons into the upper laser-level for the three-level laser system to work. The number density of electrons required for the operation of the four-level laser system is clearly much lower, due to the energy position of the lower laser-level.

The Bernard and Duraffourg condition

The Bernard and Duraffourg condition tells us something about the electron and hole densities that must be present in a direct band-gap semiconductor if we want to get LASING. An understanding of this condition requires us to handle a new concept, *quasi-Fermi-levels*. We are used to the *ordinary* Fermi-level for a semiconductor in thermal equilibrium. Under these conditions the holes and electrons obey $p \cdot n = n_i^2$. One Fermi-level is sufficient for us to calculate the carrier concentration of electrons in the conduction band, and holes in the valence band, in thermal equilibrium. However, once we get into LASER conditions we inject lots of minority carriers across the diode junction and $p \cdot n = n_i^2$ no longer holds. One Fermi-level is not sufficient to give us the band carrier concentrations. This is where the quasi-Fermi-level concept comes in. If you look at Figure A4.1 you'll see a quasi-Fermi-level for electrons in the conduction band E_{FC}, and a quasi-Fermi-level for holes in the valence band E_{FV}, for a region of direct band-gap semiconductor in which minority carriers are injected. Using these quasi-Fermi-levels, we can write down equations that look very similar to the Fermi–Dirac equation we've met before, which tell us the occupation probabilities of carriers as a function of energy. For example, the probability f_C that an electron has an energy E_C is given by

$$f_C(E_C) = \frac{1}{\exp[(E_C - E_{FC})/kT] + 1} \qquad (A4.1)$$

where

E_{FC} is the quasi-Fermi-level for electrons in the conduction band,
E_C is the electron energy.

The electron energies are measured *upwards* from the conduction band edge and are *positive* into the conduction band. We get a similar equation for the hole occupation probability:

$$f_V(E_V) = \frac{1}{\exp[(E_V - E_{FV})/kT] + 1} \qquad (A4.2)$$

where $f_V(E_V)$ is the hole occupation probability at an energy of E_V.

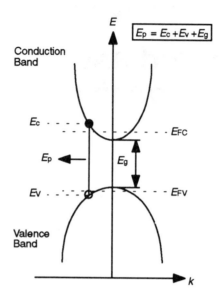

Figure A4.1   Bernard and Duraffourg condition.

The hole energies are measured *positive downwards* into the valence band from the valence band edge.

Figure A4.1 shows these various energies on an energy–momentum diagram.

I'll spell out in words the result we're going to prove, then I'll get on with the maths. We're going to show that for a direct-gap semiconductor to show *net stimulated emission*, the separation of the quasi-Fermi-levels must exceed the energy of the photon that is being emitted.

Refer again to Figure A4.1. See the photon energy E_p is given by

$$E_p = hv = E_C + E_V + E_g \tag{A4.3}$$

where E_g is the band-gap energy of the direct-gap semiconductor.

We now calculate the photon absorption and stimulated emission rates, and see what conditions are required for stimulated emission to exceed absorption. If we are to *absorb* a photon, then there must be an electron at energy E_V to absorb the photon energy, *and* no electron at energy E_C to give the photoexcited electron a place to go to. If there is an electron at energy E_V in the valence band, then *there is no hole at this energy*, and the hole occupation probability is $(1 - f_V)$. There is no electron at energy E_C in the conduction band and so the electron occupation probability at this energy is $(1 - f_C)$. The photon absorption rate is going to be *proportional* to the product of these occupation probabilities and the density of photons at energy E_p. The photon absorption rate will therefore be *equal* to

$$R_{ABS} = B(1 - f_C)(1 - f_V)\rho(E_p) \tag{A4.4}$$

where

R_{ABS} is the photon absorption rate,
$\rho(E_p)$ is the photon density at energy E_p,
B is the Einstein B (radiation) coefficient.

The analysis is very similar to finding the stimulated emission rate. In order to get stimulated emission at energy E, we need an electron to occupy energy E_C, and an empty electron state (a hole) for the stimulated electron to go to at energy E_V. The occupation probabilities are therefore f_C and f_V respectively. The stimulated emission rate will be given by

$$R_{STIM} = Bf_Cf_V\rho(E_p) \tag{A4.5}$$

If we want *net* stimulated emission (in order to make a LASER) then,

$$R_{STIM} > R_{ABS} \tag{A4.6}$$

Using equations (A4.4) and (A4.5) this means that

$$f_C + f_V > 1 \tag{A4.7}$$

Now substitute using equations (A4.1) and (A4.2) to find out what this condition means in terms of energy. You should get

$$E_{FC} + F_{FV} > E_C + E_F \tag{A4.8}$$

Finally, if you add the band-gap energy E_g to both sides of equation (A4.8) you get

$$E_{FC} + F_{FV} + E_g > E_p \tag{A4.9}$$

where E_p is the photon energy.

So, equation (A4.9) says that for net stimulated emission, the separation of the quasi-Fermi-levels must exceed the emitted photon energy.

This has an interesting application. Since the *minimum* photon energy is going to be the band-gap energy, then *at least one of the quasi-Fermi-levels must be in the band*! This condition is helped by having at least one side of the LASER junction very-heavily (degeneratively) doped.

Some useful fundamental constants

Symbol	Quantity	Value
c	Velocity of light *in-vacuo*	$2.997\,924 \times 10^{8}\,\mathrm{m\,s^{-1}}$
μ_0	Permeability of free space	$4\pi \times 10^{-7}\,\mathrm{H\,m^{-1}}$
ε_0	Permittivity of free space	$8.854 \times 10^{-12}\,\mathrm{F\,m^{-1}}$
q	Elementary charge	$1.602 \times 10^{-19}\,\mathrm{C}$
h	Planck's constant	$6.626 \times 10^{-34}\,\mathrm{J\,s}$
$h/2\pi$	Reduced Planck constant	$1.055 \times 10^{-34}\,\mathrm{J\,s}$
m_0	Free electron rest mass	$9.11 \times 10^{-31}\,\mathrm{kg}$
N	Avogadro's constant	$6.022 \times 10^{26}\,\mathrm{kg\,mol^{-1}}$
k	Boltzmann's constant	$1.38 \times 10^{-23}\,\mathrm{J\,K^{-1}}$
kT/q	Thermal voltage	$25.8 \times 10^{-3}\,\mathrm{V}$

Some properties of silicon

Atomic density	$5 \times 10^{22}\,\mathrm{cm^{-3}}$
Atomic mass	28.086
Atomic number	14
Atomic radius	118 pm
Band-gap energy (300 K)	1.12 eV
Band-gap transition	indirect
Boiling point	2628 K
Colour (bulk material)	steel-grey, shiny, opaque
Density	$2300\,\mathrm{kg\,m^{-3}}$
Dielectric constant	12.1 (low frequency) 11.9 (high frequency)
Effective density of states in conduction band	$2.8 \times 10^{19}\,\mathrm{cm^{-3}}$
Valence band	$1.04 \times 10^{19}\,\mathrm{cm^{-3}}$
Effective mass (electron)	$1.18\,m_0$
Effective mass (hole)	$0.59\,m_0$
Electron arrangement	1s(2) 2s(2) 2p(6) 3s(2) 3p(2)
Group (in periodic table)	4
Hardness	7 Mohs
Intrinsic carrier concentration	$1.4 \times 10^{10}\,\mathrm{cm^{-3}}$
Lattice parameter	0.543 nm
Melting point	1680 K
Mobility (electron 300 K)	$1500\,\mathrm{cm^2\,V^{-1}\,s^{-1}}$
Mobility (hole 300 K)	$458\,\mathrm{cm^2\,V^{-1}\,s^{-1}}$
Percentage in Earth's crust	26% by mass
Symbol	Si
Symmetry	cubic, diamond
Thermal conductivity	$26\,\mathrm{W\,cm^{-1}\,K^{-1}}$
Thermal diffusivity	$0.85\,\mathrm{cm^2\,s^{-1}}$
Valency	4

Further reading

I was in two minds as to whether to provide a book-list at all. The reason being that I find that other peoples' like, or dislike, of books is *very* personal. This being the case. you may find the books that *I* find enlightening are of very little use to *you* at all.

If you want *this* introductory book taken further (sufficient for all three years of a degree course), and with full, rigorous explanation, then you can't do much better than

Semiconductors and Electronic Devices by Adir Bar-Lev
Prentice Hall International
Third edition 1993

and

Solid State Electronic Devices by Ben G. Streetman.
Prentice Hall International
Third edition 1990

Streetman even manages to keep all the signs correct (charges, currents, etc.), a heroic effort that most of us cop out of by putting *magnitude* brackets around the quantities we're too lazy to think about.

Then there are two excellent books by S.M. Sze:

Semiconductor Devices Physics and Technology
Wiley 1985

Physics of Semiconductor Devices
Wiley
Second edition 1981

The second of Sze's books is a really massive tome, including just about every device you've ever heard of, and a few more besides. It is indispensable for reference and appears on most device physicists' shelves.

If you're keen on lasers, then

Lasers Principles and Applications
J. Wilson and J.F.B. Hawkes
Prentice-Hall 1987

is a nice concise book.

For LEDs I recommend

Light Emitting Diodes: An introduction
K. Gillessen and W. Schairer
Prentice-Hall International 1987

Some of the typography is of questionable quality but if you look beyond that to the content of the book, you'll find it to be excellent.

If you want some more quantum mechanics:

Fundamentals of Modern Physics
R.M. Eisberg
Wiley 1961

will take you a long way. This book is really excellent for thoroughly going over the basic principles, but it doesn't stop there. In later chapters the theory is applied to real systems. This gives you a good idea of how difficult it is to apply quantum mechanics practically (unlike the artificially convenient system we studied!).

Finally, although *this* is a Physics book, some of you out there might want to do some *real* electronics (build circuits). For those of you who want to do it properly and not start off as I did, in my opinion there is only one book to consider,

The Art of Electronics
Paul Horowitz and Winfield Hill
Cambridge
Second Edition 1989

I am ashamed to say that I have only recently bought and started to use this book. It shows me that I have *still* been doing it wrong (even though the circuits I have built have *fortuitously* worked). What doesn't it cover re basic circuit design? Nothing of consequence!

Index